IN-SITU REMEDIATION OF ARSENIC-CONTAMINATED SITES

Arsenic in the Environment

Series Editors

Jochen Bundschuh
University of Southern Queensland (USQ), Toowoomba, Australia
Royal Institute of Technology (KTH), Stockholm, Sweden

Prosun Bhattacharya
KTH-International Groundwater Arsenic Research Group, Department of Land and Water Resources Engineering, Royal Institute of Technology (KTH), Stockholm, Sweden

ISSN: 1876-6218

Volume 6

ISGSD

International Society of
Groundwater for
Sustainable Development

Cover photo

Upper photo: The Comarca Lagunera is one of the most contaminated regions where the aquifers are seriously contaminated with arsenic in Mexico. Continuous drawdowns due to groundwater mining mainly for irrigation purposes result in geochemical reaction which releases arsenic into the groundwater.

Lower photo: Photoremediation of arsenic-contaminated groundwater using arsenic-hyperaccumulator *Pteris vittata* (Chinese brake fern) in Florida, USA.

In-Situ Remediation of Arsenic-Contaminated Sites

Editors

Jochen Bundschuh

*University of Southern Queensland, Faculty of Health, Engineering and Sciences &
National Centre for Engineering in Agriculture, Toowoomba, Queensland, Australia
Royal Institute of Technology (KTH), Stockholm, Sweden*

Hartmut M. Holländer

University of Manitoba, Department of Civil Engineering, Manitoba, Canada

Lena Qiying Ma

*State Key Laboratory of Pollution Control and Resource Reuse,
School of the Environment, Nanjing University, Jiangsu China and
Soil and Water Science Department, University of Florida, Gainesville, USA*

CRC Press
Taylor & Francis Group
Boca Raton London New York

CRC Press is an imprint of the
Taylor & Francis Group, an **informa** business
A BALKEMA BOOK

Co-published by IWA Publishing
Alliance House, 12 Caxton Street, London SW1H 0QS, UK

First issued in paperback 2017

CRC Press/Balkema is an imprint of the Taylor & Francis Group, an informa business

© 2015 Taylor & Francis Group, London, UK

Typeset by MPS Limited, Chennai, India

Published by: CRC Press/Balkema
                P.O. Box 11320, 2301 EH Leiden, The Netherlands
                e-mail: Pub.NL@taylorandfrancis.com
                www.crcpress.com – www.taylorandfrancis.com

Library of Congress Cataloging-in-Publication Data

In-situ remediation of arsenic-contaminated sites / editors, Jochen Bundschuh,
   University of Southern Queensland, Faculty of Health, Engineering and Sciences &
   National Centre for Engineering in Agriculture, Toowoomba, Queensland,
   Australia Royal Institute of Technology (KTH), Stockholm, Sweden, Hartmut M.
   Holländer, University of Manitoba, Department of Civil Engineering, Manitoba,
   Canada, Lena Qiying Ma, State Key Laboratory of Pollution Control and Resource Reuse,
   School of the Environment, Nanjing University, Jiangsu China [and] Department of
   Soil and Water Science, University of Florida, Gainesville, USA.
     pages cm. — (Arsenic in the environment, ISSN 1876-6218; volume 6)
   Includes bibliographical references and index.
   ISBN 978-0-415-62085-7 (hardback) —
   ISBN 978-0-203-12017-0 (ebook PDF)
   1. Groundwater—Purification—Arsenic removal.   2. Soil remediation.   3. Arsenic
wastes—Environmental aspects.   4. In situ remediation.   I. Bundschuh, Jochen,
editor.   II. Holländer, Hartmut M., editor.   III. Ma, Lena Qiying , editor.
   TD427.A77I54   2014
   628.1'6836—dc23
                                                                    2014020023

ISBN 13: 978-1-138-74775-3 (pbk)
ISBN 13: 978-0-415-62085-7 (hbk)

About the book series

Although arsenic has been known as a 'silent toxin' since ancient times, and the contamination of drinking water resources by geogenic arsenic was described in different locations around the world long ago — e.g. in Argentina in 1917 — it was only two decades ago that it received overwhelming worldwide public attention. As a consequence of the biggest arsenic calamity in the world, which was detected more than twenty years back in West Bengal, India and other parts of southeast Asia. As a consequence, there has been an exponential rise in scientific interest that has triggered high quality research. Since then, arsenic contamination (predominantly of geogenic origin) of drinking water resources, soils, plants and air, the propagation of arsenic in the food chain, the chronic affects of arsenic ingestion by humans, and their toxicological and related public health consequences, have been described in many parts of the world, and every year, even more new countries or regions are discovered to have arsenic problems

Arsenic is found as a drinking water contaminant, in many regions all around the world, in both developing as well as industrialized countries. However, addressing the problem requires different approaches which take into account, the differential economic and social conditions in both country groups. It has been estimated that 200 million people worldwide are at risk from drinking water containing high concentrations of As, a number which is expected to further increase due to the recent lowering of the limits of arsenic concentration in drinking water to $10\,\mu g\,L^{-1}$, which has already been adopted by many countries, and some authorities are even considering decreasing this value further.

The book series 'Arsenic in the Environment' is an inter- and multidisciplinary source of information, making an effort to link the occurrence of geogenic arsenic in different environments and the potential contamination of ground- and surface water, soil and air and their effect on the human society. The series fulfills the growing interest in the worldwide arsenic issue, which is being accompanied by stronger regulations on the permissible Maximum Contaminant Levels (MCL) of arsenic in drinking water and food, which are being adopted not only by the industrialized countries, but increasingly by developing countries.

The book series covers all fields of research concerning arsenic in the environment and aims to present an integrated approach from its occurrence in rocks and mobilization into the ground- and surface water, soil and air, its transport therein, and the pathways of arsenic introduction into the food chain including uptake by humans. Human arsenic exposure, arsenic bioavailability, metabolism and toxicology are treated together with related public health effects and risk assessments in order to better manage the contaminated land and aquatic environments and to reduce human arsenic exposure. Arsenic removal technologies and other methodologies to mitigate the arsenic problem are addressed not only from the technological perspective, but also from an economic and social point of view. Only such inter- and multidisciplinary approaches, will allow case-specific selection of optimal mitigation measures for each specific arsenic problem and provide the local population with arsenic safe drinking water, food, and air.

We have an ambition to make this book series an international, multi- and interdisciplinary source of knowledge and a platform for arsenic research oriented to the direct solution of problems with considerable social impact and relevance rather than simply focusing on cutting edge and breakthrough research in physical, chemical, toxicological and medical sciences. The book series will also form a consolidated source of information on the worldwide occurrences of arsenic, which otherwise is dispersed and often hard to access. It will also have role in increasing the awareness and knowledge of the arsenic problem among administrators, policy makers and company executives and improving international and bilateral cooperation on arsenic contamination and its effects.

Consequently, we see this book series as a comprehensive information base, which includes authored or edited books from world-leading scientists on their specific field of arsenic research, but also contains volumes with selected papers from international or regional congresses or other scientific events. Further on, the abstract books of the homonymous international congress series, which we organize biannually in different parts of the world, will become part of this book series. The series will be open for any person, scientific association, society or scientific network, for the submission of new book projects. Supported by a strong multi-disciplinary editorial board, book proposals and manuscripts are peer reviewed and evaluated.

Jochen Bundschuh
Prosun Bhattacharya
(*Series Editors*)

Editorial board

Table of contents

Dedication to Arun Bilash Mukherjee, D.Sc. (†)

Arun Bilash Mukherjee, D.Sc.

Formerly Senior Research Scientist, Environmental Sciences,
Department of Biological and Environmental Sciences, University of Helsinki, Finland
1938–2013

We dedicate this volume to our colleague, friend and coworker Dr. Arun Bilash Mukherjee who was an editorial bord member of this book series *"Arsenic in the Environment"* since its beginning in 2008. Arun Bilash Mukherjee had been an active researcher and scientist working in the field of arsenic in the environment for a period of more than 13 years. Born in Faridpur, East Bengal (now Bangladesh) in the year 1938, he had his early education in Faridpur. He completed his degree of Bachelor of Science in Metallurgy from University of Calcutta (Kolkata), India and soon after he started his professional career as a metallurgist and worked in well known companies in the metal industry in Finland and in the copper industry in Finland, India, and Zaire (presently known as the Democratic Republic of Congo). His interest in metals brought him back to Finland in 1976, where he continued with higher studies starting with a M.Sc. in Process Metallurgy from the Helsinki University of Technology, Finland followed by the degree of Licentiate in Technology from the Department of Forest Products and he received Doctor of Science (D.Sc.) from the University of Helsinki, Finland in 1994.

His research interests included environmental biogeochemistry of trace elements, emission inventories, soil remediation, waste management and recycling, and fate of trace elements in coal and coal combustion by-products and groundwater arsenic problems in the developing countries. He had conducted several national and international investigations into trace elements and chemicals especially for the developing countries. Dr. Mukherjee had published approximately 80 papers, articles, and book chapters in peer reviewed journals, conference proceedings, series and symposia. He is also co-author of the book entitled *"Trace Elements from Soil to Humans"*, published by Springer-Verlag, Germany in spring 2007.

Since 2000, he shifted his research interest toward the global problem of arsenic in groundwater, its occurrence, fate and management for drinking water supplies. He has published a number of articles on groundwater arsenic in Bangladesh in collaboration with the KTH-International Groundwater Arsenic Research Group at the KTH Royal Institute of Technology, Stockholm, Sweden. Through the collaborative network of the KTH-International Groundwater Arsenic

Research Group, he has co-edited a number of books including "*Arsenic in Soil and Ground-water Environment: Biogeochemical Interactions, Health Impacts and Remediation*" published in the Elsevier Series "*Trace Elements in the Environment*" (2007); "*Groundwater and Sustainable Development: Problems, Perspectives and Challenges*" (2008); and "*Natural Arsenic in Groundwaters of Latin America—Occurrence, health impact and remediation*" (2009), the first volume of the Book Series "*Arsenic in the Environment*" published by CRC Press/Balkema, The Netherlands.

He had been active and energetic throughout his academic career. In 2011 he retired, followed by his illness and he passed away on 30th August 2013 in Helsinki, Finland.

The International Society of Groundwater for Sustainable Development will always remember his contributions towards the advancement of knowledge on arsenic in environmental systems.

List of contributors

Claes Bergqvist
Department of Botany, Stockholm University, S-106 91 Stockholm, Sweden

Max Billib
Leibniz University of Hannover, 30167 Hannover, Germany

Peter-W. Boochs
Leibniz University of Hannover, 30167 Hannover, Germany

Jochen Bundschuh
University of Southern Queensland, Faculty of Health, Engineering and Sciences & National Centre for Engineering in Agriculture, Toowoomba, Queensland, Australia

José Luis Cortina
Departamento de Ingeniería Química, Escola Tècnica Superior d'Enginyeria Industrial de Barcelona, Universitat Politècnica de Catalunya, 08028, Barcelona, Spain

António M.A. Fiúza
Universidade do Porto, Faculdade de Engenharia, Centro de Investigação em Geo-Ambiente e Recursos, 4200-465 Porto, Portugal

Maria Greger
Faculty of Applied Ecology and Agricultural Sciences, Hedmark University College, Blæstad, NO-2418 Elverum, Norway

Bodo Harazim
Federal Institute of Geosciences and Natural Resources, 30655 Hannover, Germany

Hartmut M. Holländer
Leibniz University of Hannover, 30167 Hannover, Germany, now at: University of Manitoba, Department of Civil Engineering, Winnipeg, MB R3T 5V6, Canada

Kyoung-Woong Kim
School of Environmental Science & Engineering, Gwangju Institute of Science and Technology (GIST), Gwangju 500-712, Republic of Korea

Soon-Oh Kim
Department of Earth and Environmental Science and Research Institute of Natural Science, Gyeongsang National University, Jinju 660-701, Republic of Korea

Timo Krüger
Leibniz University of Hannover, 30167 Hannover, Germany, now at: Ingenieurgesellschaft Heidt & Peters mbH, 29223 Celle, Germany

Keun-Young Lee
Decontamination/Decommissioning Technology Development Devision, Korea Atomic Energy Research Institute (KAERI), Daejeon 305-353, Republic of Korea

Marta I. Litter
Remediation Technologies Division, Environmental Chemistry Department, Chemistry Management, National Atomic Energy Commission, 1650, San Martín, Prov. de Buenos Aires, Argentina & National Scientific and Technique Research Council (CONICET) & Institute of Research and Environmental Engineering, National University of General San Martín, 1650, San Martín, Prov. de Buenos Aires, Argentina Argentina

Lena Qiying Ma
State Key Laboratory of Pollution Control and Resource Reuse, School of the Environment, Nanjing University, Jiangsu 210046, China & Department of Soil and Water Science, University of Florida, Gainesville, FL 32611, USA

Henning Prommer
Commonwealth Scientific and Research Organisation, Perth, WA 6009, Australia

Aurora Silva
REQUIMTE, Instituto Superior de Engenharia do Porto, Bernardino de Almeida, 4200-072 Porto, Portugal

Nandita Singh
CSIR-National Botanical Research Institute, Rana Pratap Marg, Lucknow 226001, India

Pankaj Kumar Srivastava
CSIR-National Botanical Research Institute, Rana Pratap Marg, Lucknow 226001, India

Shubhi Srivastava
CSIR-National Botanical Research Institute, Rana Pratap Marg, Lucknow 226001, India

Jens Stummeyer
Federal Institute of Geosciences and Natural Resources, 30655 Hannover, Germany

Rudra Deo Tripathi
CSIR-National Botanical Research Institute, Rana Pratap Marg, Lucknow 226001, India

Christos Tsakiroglou
Foundation for Research and Technology, Hellas – Institute of Chemical Engineering and High Temperature Chemical Processes, Platani, 26504 Patras, Greece

Aradhana Vaish
CSIR-National Botanical Research Institute, Rana Pratap Marg, Lucknow 226001, India

Dimitri Vlassopoulos
Anchor QEA, Portland, Oregon, USA

Ilka Wallis
Flinders University, Adelaide, SA 5001, Australia

Xin Wang
College of Resources and Environmental Science, Hunan Normal University, Changsha, Hunan 410081, China

Editors' foreword

Worldwide there are numerous sites where groundwater resources and the vadose zone including the soils at its top are contaminated with arsenic compounds at toxic levels. The presence of arsenic in water and soil is a global concern since it is classed as a carcinogen and presents a serious threat to human health. Arsenic in the environment is either from geogenic sources, locally accelerated by mining activities or originates from geogenic sources. Further, irrigation with arsenic-rich ground and surface water can result in arsenic-contaminated soils, constituting a severe health risk due to the uptake by plants and animal and therefore reaching the human food chain. This demand at many places of the world entails remediation measures and requires that arsenic is listed as an obligatory parameter for analysis of water used for drinking and irrigation purposes and soils used for food production. For mitigating arsenic contaminated aquifers and soils, it is necessary to identify and promote appropriate techniques to safeguard public water and food supply.

Remediation or site restoration is mostly done by *ex-situ* methods whereas *in-situ* methods do not receive the consideration it deserves. This is despite the fact that during the last two decades many *in-situ* methods have ben developed and fruitfully applied to many situations for cleaning up arsenic-contaminated groundwater and soils.

This book provides an introduction, the scientific and technological background, showcases experiences obtained so far and gives future perspectives of potentials and further need for R&D for *in-situ* arsenic remediation technologies for soils, water and groundwater at geogenic and anthropogenic contaminated sites.

We hope that this book can be used by graduate and postgraduate students and researchers in the field of environmental sciences and engineering, and hydrogeochemistry as well as researchers, engineers, environmental scientists and chemists, toxicologists, medical scientists and even for the general public seeking an in-depth view of remediation technologies for arsenic-contaminated sites. This book aims to bring awareness, among administrators, policy makers and company executives, on *in-situ* remediation technologies at sites contaminated by arsenic and to improve international cooperation.

The given case studies presented *in-situ* technologies for arsenic-contaminated sites covering arsenate and arsenite but also about organic arsenic compounds. The book also covers geochemical, microbiological and plant ecological aspect for arsenic remediation. Finally, the book includes cost effectiveness determinations and attempts for numerical modeling for the remediation and its long-term sustainability. Chapter 1 provides details of the existing *in-situ* technologies for arsenic removal or immobilization of arsenic from groundwater. It particularly shows the advantages of technologies using nanotechnology including combinations of different nanosized compounds. Chapter 2 shows how numerical modeling can be used to select one or a combination of *in-situ* technology(ies), most suitable option for specific-site conditions, to design and perform the remediation process in an optimal way and to monitor the groundwater quality during the remediation process and afterwards. The other chapters focus on *in-situ* remediation of arsenic in the vadose zone including soils. Chapter 3 provides details on different phytostabilization technologies and discusses the suitability of different options for arsenic removal considering different site-specific conditions. In Chapter 4 recent advances and innovative approaches of phytoremediation of soils contaminated with arsenic are given whereas Chapter 5 discusses the option of electrokinetic arsenic remediation. Chapter 6 covers the topic and options of microbial *in-situ* remediation of arsenic-contaminated soils. The Chapter 7 provides a case study showcasing the experiences of *in-situ* remediation of groundwater by immobilization of arsenic using a geochemical approach.

The main goal of the book is to focus attention of all affected parties worldwide on arsenic problems and to present some challenges for safe water and food production in order to promote appropriate actions in efficient, innovative directions. We hope that this book will be useful for environmental scientists and engineers in both academia and industries and for government and regulatory bodies dealing with arsenic contamination issues by providing an opportunity to acquire relevant scientific information and experiences in "*In-Situ Remediation of Arsenic-Contaminated Sites*".

Jochen Bundschuh
Hartmut M. Holländer
Lena Qiying Ma
(Editors)

Acknowledgements

The editors of this book would like to take this opportunity to acknowledge our colleagues D. van Halem (The Netherlands), W. Kujawksi (Poland), O. Sracek (Czech Republic), F. Wagner (Germany), Nandita Singh (India), Shiny Methews (USA), Maria Greger (Sweden), Gina Kertulis-Tartar (USA), Vahid Ouhadi (Canada), Kitae Baek (Korea), Giombattista Traina (Italy), Piyasa Ghosh (USA), Bala Rathinasabapathi (USA), Eduardo Moreno (Spain), Zhiqing Lin (USA), Azizur Rahman (Australia), and for their time-consuming efforts to review the manuscripts of the chapters for this volume. Their efforts with high quality review of the manuscripts contributed significantly to keep the high scientific quality of this volume. We particularly thank the editorial board members of the book series "Arsenic in the Environment" for the final editorial handling of the manuscripts. We wish to express our sincere thanks to the authors who contributed to this book. Should we have missed to mention the names of any other reviewer, we humbly apologize—and this was certainly unintentional.

Lastly, the editors thank Janjaap Blom and his team of CRC Press/Balkema, for their patience and skill for the final production of this volume.

About the editors

Jochen Bundschuh (1960, Germany), finished his PhD on numerical modeling of heat transport in aquifers in Tübingen in 1990. He is working in geothermics, subsurface and surface hydrology and integrated water resources management, and connected disciplines. From 1993 to 1999 he served as an expert for the German Agency of Technical Cooperation (GTZ) and as a long-term professor for the DAAD (German Academic Exchange Service) in Argentine. From 2001 to 2008 he worked within the framework of the German governmental cooperation (Integrated Expert Program of CIM; GTZ/BA) as adviser in mission to Costa Rica at the Instituto Costarricense de Electricidad (ICE). Here, he assisted the country in evaluation and development of its huge low-enthalpy geothermal resources for power generation. Since 2005, he is an affiliate professor of the Royal Institute of Technology, Stockholm, Sweden. In 2006, he was elected Vice-President of the International Society of Groundwater for Sustainable Development ISGSD. From 2009–2011 he was visiting professor at the Department of Earth Sciences at the National Cheng Kung University, Tainan, Taiwan. By the end of 2011 he was appointed as professor in hydrogeology at the University of Southern Queensland, Toowoomba, Australia where he leads a working group of 26 researchers working on the wide field of water resources and low/middle enthalpy geothermal resources, water and wastewater treatment and sustainable and renewable energy resources (http://www.ncea.org.au/groundwater). In November 2012, Prof. Bundschuh was appointed as president of the newly established Australian Chapter of the International Medical Geology Association (IMGA).

Dr. Bundschuh is author of the books "Low-Enthalpy Geothermal Resources for Power Generation" (2008) (Taylor & Francis/CRC Press) and "Introduction to the Numerical Modeling of Groundwater and Geothermal Systems: Fundamentals of Mass, Energy and Solute Transport in Poroelastic Rocks". He is editor of the books "Geothermal Energy Resources for Developing Countries" (2002), "Natural Arsenic in Groundwater" (2005), and the two-volume monograph "Central America: Geology, Resources and Hazards" (2007), "Groundwater for Sustainable Development" (2008), "Natural Arsenic in Groundwater of Latin America (2008). Dr. Bundschuh is editor of the book series "Multiphysics Modeling", "Arsenic in the Environment", and "Sustainable Energy Developments" (all CRC Press/Taylor & Francis).

Dr. Holländer is a civil engineer specialized in numerical groundwater modeling. He was awarded his Ph.D. in 2005. After his post-doctoral position at the Commonwealth Scientific and Industrial Organisation (CSIRO), Adelaide, Australia, he joined the Brandenburg University of Technology (BTU) Cottbus, Germany in 2008. He served from 2010 to 2012 as a research scientist in the State Authority of Mining, Energy and Geology, Hanover, Germany before he joined in 2013 as a research associate and adjunct professor the University of Manitoba, Canada.

Dr. Holländer's research program focuses on numerical studies on heat transport problems related to geothermal energy, density-driven flow, and groundwater contamination. Additionally, he conducts laboratory experiments on remediations and tests the methods in the field. He covers the undergraduate and graduate courses of Groundwater Hydrology, Groundwater Contamination, and Groundwater and Solute Transport Modelling at the University of Manitoba.

Lena Q. Ma is a Professor in the Soil and Water Science Department at the University of Florida. She earned her B.S. degree in Soil Science from Shenyang Agricultural University in 1985. She obtained her M.S. and Ph.D degrees from Colorado State University in 1989 and 1991. After spending three years as a post-doctoral scientist at the Ohio State University, she joined the University of Florida as an assistant professor in 1994. She was promoted to an associate professor in 1999 and a professor in 2003.

Dr. Ma's research program focuses on environmental soil chemistry with an emphasis on bio-geochemistry of trace metals. She conducts basic and applied research on soil contamination and remediation especially phytoremediation. She teaches both undergraduate and graduate courses including Introductory Soil, Soil Contamination and Remediation, Biogeochemistry of Trace Metals, and Graduate Seminar.

Dr. Ma's significant contributions to both basic and applied science are recognized nationally and internationally. She received the Discovery 2001 Award from the Royal Geographical Society and Discovery Networks Europe. She is the recipient of 2002 Gamma Sigma Delta Junior Faculty Award at University of Florida, 2003 Sigma Xi Junior Faculty Research Award at University of Florida and 2004 USDA Secretary's Honor Award. She was elected a fellow of American Society of Agronomy in 2002, American Society of Soil Science in 2003 and American Association for the Advancement of Science in 2012. Professor Ma published over 200 refereed journal articles and book chapters, with citation as high as 7,322 and h-index of 46.

CHAPTER 1

In-situ technologies for groundwater treatment: the case of arsenic

Marta I. Litter, José Luis Cortina, António M.A. Fiúza,
Aurora Futuro & Christos Tsakiroglou

1.1 INTRODUCTION: *IN-SITU* TECHNOLOGIES FOR GROUNDWATER TREATMENT

Conventional technologies for treating contaminated groundwater such as pump-and-treat systems have several disadvantages due to its high cost, especially when the operation is long lasting and it becomes impossible to decrease the concentration below the maximum allowable limit (Benner *et al.*, 1999; Harter, 2003; Spira *et al.*, 2006). Thus, new *in-situ* technologies are under development, such as bioremediation, permeable reactive barriers (PRBs), *in-situ* chemical oxidation, multiphase extraction, natural attenuation, electrokinetics (EK), etc. Among all these methods, PRBs are considered as the most promising ones (Bhumbla *et al.*, 1994; Gavaskar *et al.*, 1998; Gu *et al.*, 1999; Sacre *et al.*, 1997; Waybrant *et al.*, 1998).

In-situ remediation of contaminated soils and aquifers has several advantages compared to *ex-situ* techniques (Eweis *et al.*, 1998):

- the technology is environmental friendly in terms of reducing emission of volatile organic compounds (VOCs) to the atmosphere compared to excavation, transport and *ex-situ* treatment of the contaminated soil which, by trucks and excavators, contribute to the emission of greenhouse gases (e.g., VOCs and CO_2), and hence to global warming;
- the *in-situ* remediation may also be applied if the contamination is localized under roads/constructions or other places that cannot be excavated;
- in most cases, the total cost of remediation is less using *in-situ* methods in comparison to excavation and treatment;
- when extracting contaminated groundwater and treating it above ground (*ex-situ* pump-and-treat) by a variety of processes (e.g., air stripping, carbon adsorption, bioreactors, chemical precipitation, etc.) highly contaminated wastes may be produced and have to be disposed in landfills.

1.2 PERMEABLE REACTIVE BARRIERS

Permeable reactive barriers (PRBs) are a semipassive *in-situ* treatment, which uses a solid reactant that promotes chemical or biochemical reactions or sorption processes in order to transform or immobilize pollutants. The technology consists essentially of the construction of a permeable barrier that intersects the contamination plume, as shown in Figure 1.1. The barrier should allow a sufficient residence time of the pollutant to allow the occurrence of the reactions with an acceptable yield (Blowes *et al.*, 1996; Burghardt *et al.*, 2007; Ludwig *et al.*, 2009; Wilkin *et al.*, 2009).

Drilling techniques are normally used to replace the aquifer rock by the reactive material. The barrier has to be built perpendicularly to the groundwater flow, in the form of a wall. The materials need to have a high hydraulic conductivity and should intercept the flow; normally, they consist of a mixture of an inert material with a solid reagent, able to transform the pollutant

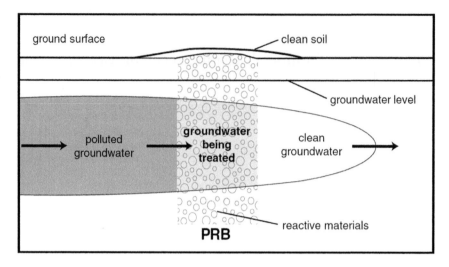

Figure 1.1.   Conceptual structure of a PRB (USEPA, 2001).

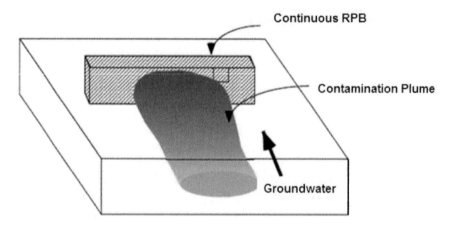

Figure 1.2.   Permeable reactive barrier. Continuous configuration (USEPA, 1998).

into an innocuous form. Materials used in the construction should obey to the following criteria (Bhumbla *et al.*, 1994; Gubert *et al.*, 2004; Younger *et al.*, 2002):

- to have enough reactivity in order to reduce the aqueous concentration of the pollutant;
- to have enough permeability to allow diverse type of waters joining the normal flux of water (in the order of 1 m per day);
- to have capacity of keeping the permeability for a long time (in order of several years);
- to have acceptable costs.

The purpose of the barrier design is to allow the complete capture of the contamination plume using the minimum of the reactive material. There are two basic types of configuration: the *continuous configuration* (Fig. 1.2) and the *funnel-and-gate configuration* (Fig. 1.3). Choosing the correct configuration depends on the size of the plume, accessibility and groundwater pattern.

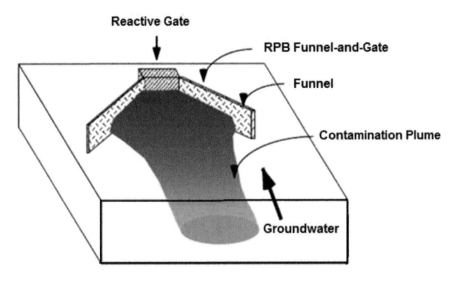

Figure 1.3. Permeable reactive barrier. Funnel-and gate configuration (USEPA, 1998).

Both configurations have been used for extensions up to 300 m but, as they require excavation, they are limited to depths up to 15–20 m.

The funnel-and-gate configuration uses classic impermeable barriers disposed as a funnel to direct the plume to the gate constituted by the PRB. The pattern flow of the groundwater is more altered with this system compared with that using the continuous configuration. In both configurations, the permeability of the gate should be higher than that of the aquifer, to avoid diversions of the groundwater around the reactive gate. PRBs are especially attractive for groundwater remediation because they conserve the water energy, and are potentially more inexpensive than the conventional pump-and-treat remediation due to lower operational and maintenance costs. Another advantage is that the reactants are used *in-situ*, thus avoiding the need for large installations and equipment on the surface.

When evaluating the suitability of a reactive medium it is necessary to account for its capacity to transform the pollutant with an appropriate kinetics, keeping an adequate permeability and reactivity during long times, and releasing only environmentally acceptable compounds as by-products.

The main processes that control the immobilization and transformation of the pollutants in the barrier include sorption on the reactive medium and precipitation, chemical reaction and biogenic reactions (Diels *et al.*, 2003). The most usual mechanism for non-polar organic compounds is sorption because of their hydrophobicity (Scherer *et al.*, 2000).

On the other hand, metals are usually adsorbed through electrostatic attraction or through a superficial complexation reaction. The suitability of sorbent materials for PRBs depends mainly on the strength of the sorbed complex and on the capacity of the material to sorb a specific pollutant. These materials have also the advantage of not releasing adverse chemicals to the groundwater, but their efficiency generally depends also on the groundwater geochemistry (e.g., pH and major anions and cations). Furthermore, metals can be immobilized increasing pH or adding an excess of ions to form an insoluble mineral phase. Thus, the metal precipitation process is a combination of a transformation process followed by an immobilization process (Chen *et al.*, 1997; Ma *et al.*, 1994).

As both sorption and precipitation are normally reversible processes, it may be necessary to eliminate the reactive materials and the accumulated products, depending on the stability of the immobilized components and the geochemistry of the groundwater.

In these terms, four different barrier types may be defined, taking into consideration the involved physical, chemical and biological processes:

- *Type 1*: Precipitation and acidity control: calcite and mixtures of calcite with siliceous gravel or other similar materials with adequate porosity.
- *Type 2*: Chemical reduction with acidity control and sulfide precipitation: calcite and zerovalent iron.
- *Type 3*: Biological reduction of sulfates with acidity control and precipitation of sulfides: a source of organic material (activated sludge, compost, wood chips), a bacterial source (sludge from anaerobic areas of local rivers and creeks) and an acidity neutralizer as for instance limestone lime (Blowes *et al.*, 1996; Gubert *et al.*, 2004).
- *Type 4*: Chemical and biological reduction of sulfate with acidity control and precipitation of sulfides: the composition of this barrier is the same as the previous one, but elemental iron is added to increase the capacity of sulfate reduction (Hammack *et al.*, 1994; Schneider *et al.*, 2001).

A correct design of a PRB implies the need to characterize the physical and chemical processes that regulate the acidity of the waters, as well as the elimination of metallic and nonmetallic species and the hydrodynamic features of these materials. For this, experiments in columns that simulate the behavior of the PRB have to be made (Hammack *et al.*, 1994). The description of the behavior of the barrier material at laboratory scale is of vital importance for a correct design and to preview its alterations with time. This can be done by incorporation of the fundamental parameters obtained experimentally using simulation models that incorporate the transport through a porous media and the chemical reactions that occur between the media and the solutes in water (Benner *et al.*, 1999; Bolzicco *et al.*, 2001).

The transformation of a pollutant in a less hazardous form through irreversible reactions does not necessarily require the elimination of the reactive medium, unless the reactivity decreases or becomes obstructed. An example of this type of transformation in a reactive barrier is an irreversible redox reaction where the pollutant is reduced or oxidized; the media may provide directly electrons for the reaction or may indirectly stimulate microorganisms to mediate the electron transfer either supplying an electron acceptor (e.g., oxygen) or an electron donor (e.g., a carbon source). In order to be effective, the electron transfer between the medium and the pollutant must be thermodynamically favorable and easy in kinetic terms (Morrison *et al.*, 2006).

Microorganisms frequently mediate redox reactions where the pollutants are either in a reduced form (e.g., petroleum hydrocarbons) or in an oxidized form (e.g., chlorinated solvents or nitrates), using the degradation of the pollutants as metabolic nest to obtain energy and material for cellular synthesis (Alvarez, 2000).

Four types of barriers are commonly used: with a sorbent, with zerovalent iron (ZVI), iron scraps and composite barriers with both organic material and elemental iron, as will be indicated in Section 1.3.2.

The design of a reactive barrier takes into account several factors. These include the reaction velocity for a specific pollutant concentration by unit of mass of reactive medium or by surface area, and the geochemistry and hydrology of the system. These factors affect the residence time of the contaminated water in the barrier necessary to reach the target pollutant concentrations. The capacity to manipulate some of these factors in predominantly passive terms would keep the cost-efficiency ratio, allowing a more flexible design and more confidence to achieve the elimination of the pollutants (Burghardt *et al.*, 2007).

The main pollutant groups studied for elimination through PRBs are the halogenated aliphatic organics (trichloroethylene, tetrachloroethylene and carbon tetrachloride), heavy metals and metalloids (hexavalent chromium, lead, molybdenum, arsenic and cadmium) and radionuclides (Blowes *et al.*, 2000; Chen *et al.*, 1997; Gotpagar *et al.*, 1997; Gu *et al.*, 1999).

1.3   REMOVAL OF ARSENIC FROM GROUNDWATER USING REACTIVE GEOCHEMICAL BARRIERS

1.3.1   *General*

For some conditions of soils and when the emission sources are not very diffuse, it is possible to eliminate As using PRBs. As said before, according to the redox conditions of the water, As can occur as arsenite or arsenate anions. It is possible to use PRBs acting according to two possible mechanisms (Blowes *et al.*, 2000; Lackovic *et al.*, 1999; Gubert *et al.*, 2004; Younger *et al.*, 2002):

- Adsorption and/or coprecipitation of the anionic As species. In this case, it is possible to use mixtures of low cost metallic oxides, such as iron oxides. Within the barrier, these metallic oxides are the minor component (10%), mixed up with silica (50%) and calcite (30–50%).
- A mechanism inherent to the used material, allowing the reduction or formation of a solid phase, such as As(0) or arsenic sulfides, depending on the presence of sulfur within the barrier. In this case, metallic iron is the active element (10%), and the remaining material was the one stated before. The active barrier would take advantage of the redox properties of the As(III)/As(V) system, as it was studied for other metals and metalloids with high oxidation number such as Cr(VI), Mo(VI), U(VI), Se(VI) or organic compounds in oxidized forms (Bianchi-Mosquera *et al.*, 1994; Blowes *et al.*, 1997; Deng and Hu, 2001; Fryar *et al.*, 1994; Guillham and O'Hannesin, 1992; Joshi and Chaudhuri, 1996; Ptacek *et al.*, 1994). Some materials have been tested with high success such as siderite ($FeCO_3$), pyrite (FeS_2) and Fe(0) in the zerovalent form of granular fillings [Fe(s)]. Generally, the reactions that have been considered responsible for the process are the reduction of As(III) and As(V) to the zerovalent form or to a sulfide, while Fe(II) is oxidized to an oxide-hydroxide form that can originate coprecipitation mechanisms and/or As adsorption. These mechanisms are not completely clear.

1.3.2   *PRB types for treating arsenic in groundwater*

Several materials have been used in PRBs for treating As in groundwater. The most common are: (i) elemental iron, (ii) slag from iron works, (iii) sorbent materials such as mixtures of iron hydroxides and activated alumina, (iv) multifunctional barriers, either multiple or composed, consisting in a first barrier of compost or another organic material that promotes the microbial reduction of sulfates, followed by a second barrier or another sorbent material.

1.3.2.1   *PRBs with Fe(0)*
Most of this type of barriers uses metallic iron [Fe(0)] as reactive medium to convert the pollutants into non-toxic species or species with low mobility. These barriers take advantage of oxidation-reduction processes where the pollutant is reduced and the medium is oxidized. Zerovalent metals such as iron, tin and zinc, that are moderately strong reducing agents, have been used as reactive media (Powell *et al.*, 1995). Iron is the most studied one, and its efficiency was proved for several types of pollutants.

Technologies based on zerovalent iron (ZVI) are considered by the US Environmental Protection Agency as a valuable method to eliminate traces of organic pollutants, and it has also been considered as potentially adequate to eliminate As and metallic species. Its efficiency as a reducing agent for several organic and inorganic pollutants was also demonstrated. Among the organics, the following can be mentioned: aliphatic chlorinated compounds, nitroaromatics, some pesticides and azo-dyes; among the inorganics, metallic species of high valence such as chromium(VI), uranium(VI), technetium(VII), mercury(II), molybdenum(VI) and inorganic non-metallic anions such as nitrate, nitrite, selenate, selenite, arsenite and arsenate, among others (Alvarez, 2000; Deng and Hu, 2001; Guillham and O'Hannesin, 1992; Lien and Wilkin, 2005). The transformation process is a surface reaction that requires an intimate contact between the reactive medium and the pollutant. Thus, the global process must be considered as a series of basic physical and chemical processes:

- mass transport of dissolved pollutant from the solution to the metallic surface;
- sorption of the pollutant onto the metal surface;
- electron transfer from the metal surface to the pollutant;
- desorption of the pollutant from the metal surface.

Any of these processes may be the determining step of the pollutant reduction rate, and this depends on each specific pollutant.

Fe(0) is efficient for As(III) and As(V) removal; an advantage of ZVI is that it is affordable and non-toxic (Cundy *et al.*, 2008; Su and Puls, 2003; Wilkin *et al.*, 2009; Zouboulis and Katsoyiannis, 2005). Although not totally understood, the main mechanism seems to be superficial complexation and precipitation on surface and adsorption. The involved reaction can be depicted as follows: when iron is oxidized, FeOOH is produced at the surface, having the capacity to adsorb metals and metalloids such as As (Nikolaidis *et al.*, 2003). As Fe(0) is very efficient as reducing agent, it can remove both inorganic and organic As. ZVI is especially efficient for As removal at low pH in waters with high sulfide concentrations. In spite of the fact that the reduction capacity of elemental iron decreases significantly at neutral pH, the hydroxylated species formed on the Fe(0) surface are effective sites for adsorption of As(III) and As(V) at neutral and alkaline pH. Adsorption of As onto Fe(0) is widely influenced by the following anions, mentioned in decreasing influence order: phosphate, silicate, chromate and molybdate, followed by carbonate and nitrate and, finally, borate and sulfate.

In the transformation process, iron suffers several oxidation reactions (Powell and Puls, 1997):

$$Fe(0) \rightarrow Fe^{2+} + 2e^- \qquad E^0 = -0.447\,V \qquad (1.1)$$

$$Fe(0) \rightarrow Fe^{3+} + 3e^- \qquad E^0 = -0.037\,V \qquad (1.2)$$

$$Fe^{2+} \rightarrow Fe^{3+} + e^- \qquad E^0 = 0.771\,V \qquad (1.3)$$

In the absence of strong oxidants, there are two reduction half-reactions that, together with iron oxidation, originate spontaneous oxidation in water; in aerobic conditions, the preferential oxidant is oxygen:

$$O_2 + 2H_2O + 4e^- \rightarrow 4OH^- \qquad E^0 = 0.401\,V \qquad (1.4)$$

$$2Fe(0) + O_2 + 2H_2O \rightarrow 2Fe^{2+} + 4OH^- \qquad (1.5)$$

$$2Fe(0) + O_2 + 4H^+ \rightarrow 2Fe^{2+} + 2H_2O \qquad (1.6)$$

while, in anaerobic conditions, water acts as the oxidant:

$$2H_2O + 2e^- \rightarrow H_2(g) + 2OH^- \qquad (1.7)$$

$$Fe(0) + 2H_2O \rightarrow Fe^{2+} + H_2(g) + 2OH^- \qquad (1.8)$$

Fe(II) oxidation follows:

$$4Fe^{2+} + O_2(g) + 10H_2O \rightarrow 4Fe(OH)_3(s) + 8H^+ \qquad (1.9)$$

The common ions of these solutions may affect the efficiency of Fe(0) barriers to remove As and the competitive coprecipitation-sorption processes through the formation of the following phases (Su and Puls, 2003):

$$3Fe^{2+} + Fe^{3+} + Cl^- + 8H_2O \rightleftarrows Fe_4(OH)_8Cl(s) + 8H^+ \qquad (1.10)$$

$$4Fe^{2+} + 2Fe^{3+} + SO_4^{2-} + 12H_2O \rightleftarrows Fe_6(OH)_{12}SO_4(s) + 12H^+ \qquad (1.11)$$

$$4Fe^{2+} + 2Fe^{3+} + CO_3^{2-} + 12H_2O \rightleftarrows Fe_6(OH)_{12}CO_3(s) + 12H^+ \qquad (1.12)$$

Both the reactions mediated by oxygen or by water result in a pH increase, although the effect is more pronounced in anaerobic conditions because corrosion occurs faster.

From this description, it can be concluded that there are three main reducing agents in a water-iron system: metallic iron, ferrous iron and hydrogen resulting from the corrosion (in anaerobic conditions). The relative intensity of each one in the process depends on the compound to be reduced. The corrosion of Fe(0) particles or aggregates produces Fe(II), Fe(III) and OH^- ions, which promote in turn the precipitation of the Fe(II) hydroxide [$Fe(OH)_2(s)$] and several Fe(II/III) oxyhydroxides and $Fe(OH)_3(s)$. These precipitation reactions may favor both As coprecipitation with iron minerals as well as sorption of As on the corroded iron particles, thus contributing to the elimination of As from the solution (ZVI) (Beak and Wilkin, 2009; Burghardt *et al.*, 2007; Su and Puls, 2003). This mechanism is considered as a complex process, not completely known in terms of the involved reactions, especially at the surface of the precipitated solids, whose chemical composition is not known, and dependent on pH and on the aqueous phase composition. The knowledge of the involved reactions, if possible, would allow a better design of the As removal processes. So far, most studies have been carried at laboratory scale with zerovalent iron materials for As removal (Gu *et al.*, 1999; Morgada *et al.*, 2009; Su and Puls, 2003; Triszcz *et al.*, 2009) and only a few experiments have been reported at a full barrier scale.

When the water reacts with Fe(0), besides the increase in pH, the redox potential decreases and the oxygen concentration increases. The higher pH favors the precipitation of calcium and iron carbonates, as well as the insoluble metallic hydroxides. The decrease in potential originates the reduction of metals and metalloids. Finally, the increase in the oxygen partial pressure supports the activity of chemotrophic microorganisms that use hydrogen as energy source, especially sulfate and iron reducing bacteria. As(V) anion in water bounds iron originating its oxidation to ferrous iron through aerobic or anaerobic mechanisms.

Kinetics is fast: McRae (1999) observed As(V) removal from concentrations from 1000 to less than $3 \, \mu g \, L^{-1}$ in about two hours. Kinetics is also very fast with mixtures of As(III) and As(V). Mineralogical studies show that As(V) reduces and coprecipitates with iron, which transforms into goethite over the elemental iron particles.

The reduction yield depends on the superficial area of iron; in many cases, it is possible to observe a linear relationship between both parameters (O'Hannesin and Gillham, 1998; Puls *et al.*, 1999), even though the yield seems to stabilize for high superficial areas (Johnson, 1996; CL:AIRE, 2001).

Bang *et al.* (2005) refer that As removal is dramatically affected by DO concentration and the pH, once high DO concentrations and low pH increase iron corrosion (McRae, 1999). They found that under oxic conditions As(V) removal is considerably faster than that of As(III). At pH 6, more than 99.8% As(V) was removed compared to 82.6% As(III) after 9 h contact time. When DO was eliminated by purging with nitrogen, total As removal was below 10%.

1.3.2.2 *Barriers with iron slag*

Baker *et al.* (1998) used slag composed of a mixture of iron oxides, calcium oxides and limestone as reactive medium. This medium proved its ability to remove not only As(V) but also mixtures of As(III) and As(V) from concentrations up to $1000 \, \mu g \, L^{-1}$ to concentrations below $3 \, \mu g \, L^{-1}$.

McRae (1999) tested mixtures of slag from steel production (BOFS – Basic Oxygen Furnace Slag) as material for possible use in PRBs. The material promotes oxidation of As(III) to As(V), and it is used together with activated alumina, which adsorbs As in both oxidation forms. The tested mixture had 10% of slag and 20% of activated alumina mixed with limestone and silica sand. In contrast, the use of slag without other components showed less satisfactory results, and could not be considered a viable option.

This alternative was immediately implemented in real scale. The "Dupont Site" barrier, built in Eastern Chicago in 2002, uses BOFS for the remediation of groundwater contaminated with As. The slag is rich in iron and in calcium oxyhydroxides. The system comprises two permeable barriers placed at a distance of 5 m. The reactive medium oxidizes As(III) to As(V), which is then sorbed onto the slag surface. pH increases during the process, reaching values as high as 12.

Table 1.1.   Some commercially available arsenic sorbents.

Name of the product	Company	Material type
Adsorpas	Technical University of Berlin	Granular iron hydroxide – α-FeOOH
ARM 300	BASF	Iron oxides (hematite, α-Fe$_2$O$_3$)
G2	ADI International	Modified iron; diatomites covered with iron hydroxides
SMI III	SMI	Iron/sulfur
GEH	Wasserchemie GmbH	Granular iron hydroxide; Fe(OH)$_3$ and FeOOH (akaganeite, β-FeOOH)
Bayoxide E33	Bayer AG	Iron oxide; 90% goethite (α-FeOOH)

1.3.2.3   *Barriers with mixtures of iron hydroxides and activated alumina*

GFH (granular ferric hydroxide) and GFO (granular ferric oxide) are excellent As sorbents. GFH is prepared from a solution of ferric chloride and precipitation with sodium hydroxide; the material is washed, centrifuged and granulated at low pressure (Driehaus *et al.*, 1998). Aqueous silicates interfere and reduce the removal capacity of As(V) by GFH.

Table 1.1 shows some other iron-based sorbents (IBS) available in the market. The sorption is of chemical origin and thus it is irreversible. IBS may be used either in fixed bed columns identical to those used with activated alumina (AA) or in PRBs.

The affinity of these sorbents for As under natural pH conditions is much higher than that of AA. This fact allows IBS to treat a much higher total bed volume without pH adjustment. Nevertheless, just as with AA, the best behavior of IBS is attained at low pH. In columns, the recommended operational conditions imply a residence time of five minutes and a hydraulic charge of 0.2 (m^3 min^{-1}) m^{-2}. Phosphates compete with As(V) for the sorption sites, and each increment of 0.5 mg L^{-1} above the threshold of 0.2 mg L^{-1} reduces the sorption capacity by 30%.

Ipsen *et al.* (2005a), in a research on the identification and testing of materials that could be used in PRBs, mention that the best sorbents are materials based on akaganeite/ferrhydrite (β-FeOOH/Fe$_2$O$_3 \cdot 5$H$_2$O). The commercial product GEH presented a loading capacity of 36 g As kg^{-1}, showing a better behavior than other tested materials. The minerals of the goethite type (α-FeOOH), improved by addition of titanium, showed clearly a less loading capacity.

Jang *et al.* (2007) showed that amorphous hydrous iron oxides embedded in natural diatomite (aluminum silicates) were more efficient than elemental iron for As removal.

Silva *et al.* (2009) studied hydrous iron oxides loaded on activated carbon (HFO/AC) and the commercial sorbent ARM 300 as possible As sorbent materials for application in PRBs. The last material presented the largest loading capacity, 49 ± 20 g kg^{-1}, followed by HFO, 38 ± 2 g kg^{-1}, and finally by HFO/AC, only 5.5 ± 0.5 g kg^{-1}.

No barriers based on AA or IBS have yet been constructed so far.

1.3.2.4   *Composite barriers*

The fourth type of conception for a PRB is the composite or multiple functional barrier. Using this design, a first barrier is located upstream the source of As contamination; the reactive material is organic matter, usually compost or wood chips, promoting the biological reduction of sulfates, with subsequent precipitation of metal sulfides. Downstream to the contamination source, a second conventional barrier is built, for instance, with Fe(0). As first step, the sulfides released by the organic matter can start a reductive dissolution of the As loaded into the Fe and Mn oxides and hydroxides. This causes an increase in the As concentration over a short period of the As elution and reduces the remediation time. On the other hand, the precipitation of As sulfides reduces the As emissions, thus increasing the lifetime of the downstream barrier. Precipitations occur in the form of As_2S_3 or by coprecipitation with iron sulfides. If the concentration of sulfates in the groundwater is too low, this could be increased by dissolution of gypsum. Several researchers have studied this type of potential barrier, including Gubert *et al.* (2004) and Köber *et al.* (2003; 2005).

1.4 APPLICATIONS OF PRBS

1.4.1 *Application of Montana*

In June 2005, an experimental barrier was installed near the town of Helena, Montana (USA, Wilkin *et al.*, 2006), with a length of 9 m, a thickness of 7.6 m and a variable width between 1.8 and 2.4 m. The barrier was located near an old lead smelter and was designed to treat underground water with moderate concentrations of As(III) and As(V). Groundwater at the site had become contaminated with arsenic due to leaching from the contaminated process ponds located over the shallow groundwater. The arsenic plume was approximately 450 feet wide and extended 2100 feet down gradient from the process ponds. The barrier was built in 3 days, using digging equipment modified to allow the construction of deep trenches that were filled with a pulp of a biopolymer. The reactive medium was completely composed of granular ZVI. Concentrations of As upstream the barrier exceeded $25 \, g \, L^{-1}$. From the eighty samples taken downstream the barrier, eleven exceeded $0.50 \, mg \, L^{-1}$, 62 had concentrations below $0.05 \, mg \, As \, L^{-1}$ and 24 were below the acceptable limit of $10 \, \mu g \, L^{-1}$. After 2 years of operation, the concentrations downstream the barrier were considerably lower.

1.4.2 *Application to the treatment of groundwater contaminated by acid drainage from pyrite mines*

One of the biggest problems in the mining sector is the treatment and disposal of solid waste and liquid effluents generated in the processing stages, especially when the facilities are abandoned. The best example is mining dedicated to the processing of sulfides, where the adequate management of acidic effluents generated in the process is the main objective to reduce the environmental impact of the activity. Generally, mixtures of solid and liquid waste are accumulated in controlled damps as disposal treatment. During these periods, soluble sulfide species can be oxidized by dissolved oxygen in water according to the following reaction (Blodau, 2006):

$$\text{Metallic sulfide} + \text{water} + \text{oxygen} \rightarrow \text{soluble metal} + \text{sulfate} + H^+ \quad (1.13)$$

Therefore, the water of these ponds has variable amounts of ions (Fe, Zn, Pb, Cu), nonmetallic ions [As(V), As(III)], a high acidity (pH between 1 and 2) and high sulfate contents. Among the constituents of acid mine water, As has been recognized as one of the pollutants with the greatest impact on aquatic ecosystems because of its persistence, toxicity and bioaccumulation, and toxicological effects associated (Cundy *et al.*, 2008; Morgada *et al.*, 2009; Tyrovola and Nikolaidis, 2009; Zouboulis and Katsoyiannis, 2005). In the absence of specific regulations, the same regulation criterion of drinking water ($10 \, \mu g \, As \, L^{-1}$) has been selected for the remediation process with PRBs in groundwaters.

While the case of surface run-off and drainage treatment is relatively affordable, in the case of groundwater, pump-and-treat technologies with *ex-situ* treatment and return of the treated water to the aquifer are hardly applicable due to its high cost and reduced capacity to achieve the required quality standards (Spira *et al.*, 2006). It is therefore proposed the use of PRBs, replacing the aquifer material by reactive materials to treat the pollution plume generated (Sacre, 1997; Waybrant *et al.*, 1998). Groundwater moves through the barrier treatment by natural flow or, when it is necessary to pump, wells can be installed in a way that the water contaminant passes through the reactive barrier (Burghardt *et al.*, 2007; Lackovic *et al.*, 1999).

Table 1.2. Composition of groundwater from wells near the Aznalcóllar dam used in the experiments with columns for evaluation of reactive mixtures of lime, organic matter and iron.

Composition	Water I	Water II
Ca(II) [mg L^{-1}]	360	360
Fe(II) [mg L^{-1}]	10	10
Zn(II) [mg L^{-1}]	20	20
Cd(II) [mg L^{-1}]	–	2
Cu(II) [mg L^{-1}]	–	20
Al(III) [mg L^{-1}]	10	10
As(V) [mg L^{-1}]	2	2
SO$_4^{2-}$ [mg L^{-1}]	960	960
HCO$_3^-$ [mg L^{-1}]	24	24
pH	3.0	3.0

1.4.3   *The Aznalcóllar pollution case*

On 25 April 1998, a rupture occurred in the tailings dam at the Aznalcóllar mine (SW Spain, Bolzicco *et al.*, 2001), and the Guadiamar river received a 6 hm^3 of sludge and acidic water. This event caused a major ecological catastrophe, and life in the river disappeared completely, reaching the acidic waters the entrance limits of the Doñana National Park, one of the most important wetlands in Europe. Furthermore, in aquifers near the reservoir, detectable levels of metal species, arsenic, sulfate and acidity were measured. For this problem, the best option for remediation was the use of a PRB, based on previous experience with a similar situation in the Nickel Rim mine (Dudbury, Canada), where sulfate (2000–5000 mg L^{-1}), iron (250–1300 mg L^{-1}) and slightly acidic pH (5–6) have been detected (Joshi and Chaudhuri, 1996). Placing a barrier based on calcite, organic matter and gravels, allowed the reduction of sulfate to values between 200 and 3600 mg L^{-1} and iron to values between 1 and 40 mg L^{-1}, maintaining the control of acidity near 7. According to these results, the PBR was considered a viable, efficient and economical solution (USEPA, 1998).

Considering the analysis of groundwater from wells near the Aznalcóllar pond, two different types of model waters were evaluated at laboratory scale as preliminary evaluation (Waters I and II), whose composition is shown in Table 1.2. As can be seen, waters were rich in calcium and sulfate, with an acidic pH.

Given the situation of the Guadiamar river and taking into account the composition of groundwater and the composition of the sludge dispersed, it was necessary to act on the following aspects: (i) to regulate the acidity of water, (ii) to reduce sulfates, (iii) to reduce the levels of heavy metals, especially zinc, and (iv) to be aware of the possible mobilization of As and the effect of the barrier for its abatement. In this case, As removal was not a priority because its concentrations were close to the limits set by the WHO (Gavaskar *et al.*, 2005).

The materials used in laboratory scale tests are described in Table 1.3. The used calcite was natural and had a particle size of 2 mm. Two types of vegetal compost were evaluated: one from plant debris from the area close to Aznalcóllar and another one obtained from organic matter of solid wastes and sewage sludge. The source of sulfate-reducing bacteria was obtained from river sediments of an anaerobic zone. Shavings of metallic iron used previously in reactive barriers for decontamination of organochlorinated compounds were used; shavings from machining processes of metallic pieces of cast steel were also tested.

The efficiency for As removal of the reactive mixtures containing organic matter was very high in both columns since the beginning of the experiments. The reactive material containing Fe(s)

Table 1.3. Materials used in column laboratory experiments with polluted waters from the alluvial Guadiamar aquifer (Seville, Spain).

Material	Composition	Size
Calcite	$CaCO_3(s)$	2 mm
Compost	(i) Vegetable waste	–
	(ii) Urban solid waste and sewage sludge	–
River sediment	Sulfate reducing bacteria source	
Fe aggregates	$Fe(s)$ (90%)	8–80 mesh

yielded 99.5% efficiency removal, attaining As levels below the limit of the WHO in drinking water ($10 \mu g L^{-1}$). For the reaction mixture in columns without Fe, there was an initial erratic trend of variable As concentrations (up to $190 mg L^{-1}$) during the first two months of the experiment. Then, the concentrations were maintained between 10 and $20 \mu g L^{-1}$, very close to the WHO standard. The efficiency of both materials for removing reactive metals (Zn, Cu, Cd, Al and Fe) (not shown in this text) was very high (greater than 99%). However, despite the high removal of metal species and As, a net consumption of sulfate was not detected, suggesting that the removal of As and metal ions is due to other processes not associated with precipitation as metal sulfides.

These studies concluded that oxidation products of Fe(s) are the optimal phase for removing As species. It can be concluded that organic matter used in vegetal municipal compost behaved as a poor carbon source to support the processes of sulfate reduction. However, mixtures of organic/calcite/Fe(s) were very efficient in removing metallic and As species, in this last case with measured concentrations always below the value set by WHO of $10 \mu g L^{-1}$. Both reaction mixtures are being validated in a barrier of 120 m length, 1.4 m width and an average depth of 6 m in the aquifer of Guadiamar since 1999 (Bolzicco *et al.*, 2001).

1.5   LIMITATIONS OF IRON REACTIVE BARRIERS:
       USE OF REACTIVE ZONES

Permeable reactive barriers are tested presently to treat a growing number of contaminants in groundwater, mainly organics. In fact, many places are being decontaminated with this technology. However, despite the success of these barriers, there are still significant limitations that come from physical and geochemical characteristics of the particular site (Wilkin *et al.*, 2003). Such problems arise in the case of applications to inorganic contaminants [Cr(VI), As(V)], which means that for these cases the iron barrier technology is not considered as proven for the USEPA.

The most important limitation is the lack of information on the long-term effectiveness of a large-scale process. Even though some real scale studies appear promising and suggest means for various durations in the range of decades, caution should be exercised when predicting clearance rates scale studies from laboratory data of short duration. Removal rates in real scale applications can be influenced by long processes such as aging of the reagent or decrease in permeability due to precipitation, microbiological growth or accumulation of gas, which cannot be predicted in short-term laboratory experiments (Lo *et al.*, 2007; USEPA, 1998).

In addition to the physical limitations imposed by the geology of the site and commercial excavation techniques, there are many unresolved issues concerning the process of elimination. Ideally, the contaminants are permanently immobilized or transformed into non-hazardous substances. For iron barriers, an unresolved issue is the role of ferrous iron precipitates and the surface impurities in the reduction process. In theory, these compounds may serve not only as reducers, but also as catalysts. Ferrous iron bound to the surface or complexed species can be more directly responsible for the reduction of a contaminant. Moreover, the strength of ferrous iron as a reductant could be significantly affected by the ligands present in the system, including organic matter and metal oxides that form complexes with ferrous iron (Beak and Wilkin, 2009; Herbert *et al.*, 2000; Roh *et al.*, 2001).

The technological challenge is to reduce or eliminate the problems of possible obstruction of the pores or those that can be a physical barrier for the active sites. Alternative strategies to remove oxides include ultrasound technologies and pH control. In addition, there may be microbial processes that help to alleviate the reduction in pore volume due to the formation and trapping of hydrogen gas produced by the anaerobic corrosion of iron (Phillips *et al.*, 2003). As already stated, removal of metals and metalloids through ZVI is a combination of transformation and immobilization processes, in which the species is reduced to a less soluble form. The main challenge of the method is to avoid the potential risk of remobilization due to dissolution of the compounds formed (Lo *et al.*, 2007).

Three important factors that determine the usefulness of a metallic iron material for *in-situ* decontamination have been determined (Phillips *et al.*, 2003; Wilkin *et al.*, 2003). One is the iron content, since it must at least be sufficient to stoichiometrically react with the contaminant. For many contaminated aquifers, the overall concentration is small, in the order of $mg L^{-1}$ or $\mu g L^{-1}$. However, because of the slow flow of groundwater and the division of contaminants within and outside the solid phase, a typical plume may require between 100 and 200 years to cross a certain point (Wilkin *et al.*, 2002).

There is also uncertainty about the limiting step of the elimination process. In general, rates of transformation in a reactive barrier may be controlled by transport to the surface or by the surface reaction. Interactions between abiotic and biotic processes represent a particularly difficult research challenge because of the complexity of the potential synergy or antagonistic effects. On one hand, in short-term studies, the microorganisms seem to improve the kinetics and the distribution of final products; however, the effect of microorganisms on permeability and long-term reactivity is less well known.

In addition, it is important to note that, apart from the limitations of the barrier and the medium, there are also important costs involved. Although the total cost of an iron reactive barrier is lower than a pump-and-treat system, the initial costs of installing the barrier is much higher, considering that several long-term data are still unknown and it is difficult to overcome the reluctance to the use of this technology for commercial purposes.

The second factor is how efficiently the reducing agent is used. Excavation is the most economical on-site decontamination method. The ability to reduce a greater number of moles of contaminant with the same number of moles of iron means that it is likely that a smaller volume of material barrier may be required for a specific application and, therefore, the total volume to be excavated can be reduced (Mackenzie *et al.*, 1997).

The third factor is the rate of iron corrosion, as the corrosion reactions of iron with oxygen and water are thermodynamically favored. It is important that these reactions are sufficiently slow in the time scale of the decontamination process.

Based on the existing technology in the different sites, it is doubtful that the barriers are cost-effective for the treatment of deep aquifers (deeper than 30 m), or in places geologically difficult to access. Injection of a reactive material (e.g., colloidal iron) by hydraulic fracturing, mixture in depth or injection under pressure can be an alternative that overcomes the limitations associated to the commercial excavation technologies. Thus, a remediation proposal could be the use of colloidal iron or iron nanoparticles as substitutes for iron particles, usually in the millimeter scale (3–6 mm). The conceptual change of this application is described in Figure 1.4 (Kaplan *et al.*, 1996). When the choice of treatment is the one described above, the objective is the introduction of colloidal or nanometric particles in areas of the aquifer that can capture the contamination plume. This type of applications has been recognized in the literature as reactive zones (Diels and Vanbroekhoven, 2008). This implies, as described in Figure 1.4, that it is not necessary to drill the trench in which the reactive material is introduced, but it is introduced using injection wells from the surface.

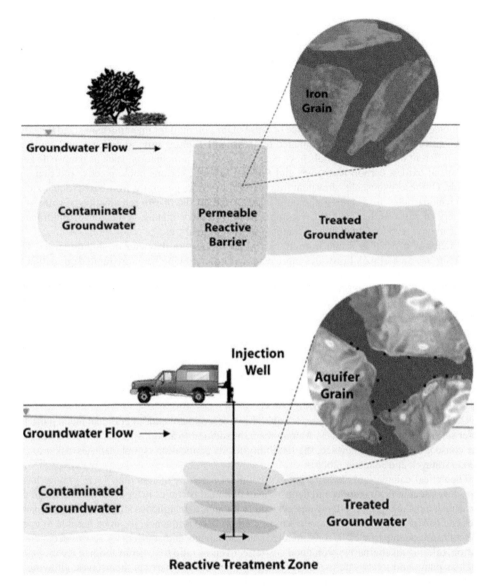

Figure 1.4. Schematic representation of changes resulting from using reactive barriers with added metallic iron with particle size in the range of 3–6 mm (upper part) *vs.* the use of nanoparticles or iron colloids (Kaplan *et al.*, 1996).

1.6 USE OF IRON NANOPARTICLES

1.6.1 *Increase of the reactivity by size decrease*

As said before, iron has a high efficiency for elimination of a large amount of organic and inorganic pollutants. However, reactivity is limited by the characteristics of the medium, the pollutant and the material. A high reactivity is important for the application in reactive barriers because the major part of the cost of this technology is in the excavation and installation of the barrier; for this reason, the higher the volume of the barrier, the higher will be its cost. Therefore, if the reactivity of the material can be increased, the necessary amount will decrease and, consequently, the final volume of the barrier. A way to improve the reactivity consists in increasing the specific surface area of the material, because of the increase of the area/volume ratio; i.e., with the same total mass of the reactive material a higher available mass for the reaction can be obtained. The amount of available surface area is among the most significant experimental variables that affect the rate of reduction of pollutants. To attain higher specific surface areas, different techniques have been developed in the case of iron and other metal nanoparticles (Ponder, 2003; Zhang, 2003).

Nanotechnology has special relevance because of the potential for injecting nanosized (reactive or adsorptive) particles into contaminated porous media such as soils, sediments, and aquifers. Many different nanoscale materials have been explored for remediation, such as nanoscale zeolites, metal oxides, carbon nanotubes and fibers, enzymes, various noble metals, and titanium dioxide. Of these, nanoscale zerovalent iron (nZVI), and more generally, iron-based (e.g., oxides, salts) nanoparticles are currently the most widely used for the *in-situ* remediation of soils from a variety of toxic pollutants (e.g., immobilization of heavy metals, reduction of chlorinated hydrocarbons, sorption/geochemical trapping of heavy metals, etc.).

Particles of nZVI may range from 10 to 100 nanometers in diameter or slightly larger. The larger surface area of nZVI provides more reactive sites, allowing for more rapid degradation of contaminants when compared to macroscale ZVI.

In addition, methods of preparation of nanoparticles are not expensive and, together with the high reactivity of the particles, they can be competitive with commercial iron particles. It is also important to take into account that the material usually used in reactive barriers is not pure iron, but a commercial product, consisting of metal wastes, mostly molten iron, or light alloy material, and it is coated by a thick oxide layer, having a lower reactivity than that of pure iron (Westall, 1986).

Due to its reduced particle size and high reactivity, metal nanoparticles can be useful in a large variety of environmental applications such as treatment of soil and sediments and decontamination of groundwater (Westall, 1986; Zhang, 2003).

Compared with conventional particles of higher size, colloidal and subcolloidal metal particles offer several potential advantages. Among them, in addition to a high specific surface area, with the consequent increased surface, the flexibility for its application is included (Gavaskar *et al.*, 2005; Tratnyek and Johnson, 2006).

Theoretical calculations indicate that for colloidal particles of less than 1 μm gravity has a very low effect in the transport and deposition of colloidal particles in a porous medium, and the Brownian movement (thermal movement) tend to dominate. In aqueous solution, iron nanoparticles can be kept suspended under a very mild stirring. In consequence, it can be feasible to inject subcolloidal metal particles in contaminated soils, sediments and aquifers for *in-situ* decontamination, offering an alternative with good cost-effectiveness ratio to the conventional technologies such as pump-and-treat systems, air stripping or even reactive barriers themselves, allowing in addition their application in aquifers that cannot be treated by other methods due to is high depth (Ponder *et al.*, 2000).

The global performance of a nanoparticle system could be potentially thousands of times better than that using commercial iron. This is especially important for the injection of iron particles in groundwaters, because to avoid obstructions it is beneficial to inject only a low amount of very reactive metal particles. Thus, in respect to its application in removal of pollutants, nanoparticles have two main applications: (i) direct injection to the polluted medium, or (ii) to be supported in some type of material for use as reactive barriers.

1.6.2   *Preparation of iron nanoparticles*

There are several forms of preparation of nanoparticles, which can be divided in two main groups: physical methods and chemical methods. In the first group, the metal nanoparticles are formed from atoms in the process of vaporization of the metal and subsequent condensation in various supports, or they are obtained through the treatment of particles of higher size in colloidal dispersions, by means of colloidal mills, ultrasound, etc. In the second group, the main method is reduction of metal ions in solution in conditions that favor the ulterior formation of small metal aggregates. The main disadvantage of the chemical synthesis in aqueous phase is the ample distribution of sizes of the metal particles and their relative low stability (Chatterjee, 2008; Ponder, 2003).

Numerous methods have been developed for the manufacture of metallic nanoparticles, including chemical vapor deposition, inert gas condensation, pulsed laser ablation, spark discharge generation, sputtering gas-aggregation, thermal decomposition, thermal reduction of oxide compounds, hydrogenation of metallic complexes and aqueous reduction of iron salts. These manufacturing methods can be considered as either "bottom up" or "top down" approaches. The former involves physical or chemical methods to construct a nanomaterial from basic building blocks, such as atoms or molecules. The latter involves physical or chemical methods to breakdown or restructure a bulk material to the nanoscale (Crane and Scott, 2012).

A simple method of synthesis of nanoparticles is in solution, from iron salts (for example, $FeCl_3 \cdot 6H_2O$, $Fe_2(SO_4)_3 \cdot 5H_2O$ or $FeCl_2 \cdot 4H_2O$), which are reduced by a reductant such as hydrazine, sodium borohydride or hydrogen, or another especial reductive medium (Chatterjee, 2008). The borohydride reduction of ferrous salts is the most widely studied method within academia (Wang and Zhang, 1997). Sodium borohydride has a reductive power adequate for several metals in normal conditions. For Fe(III) compounds:

$$Fe(H_2O)_6^{3+} + 3BH_4^- + 3H_2O \rightarrow Fe(0)(s) + 3B(OH)_3 + 10.5H_2(g) \qquad (1.14)$$

For Fe(II) compounds:

$$Fe(H_2O)_4^{2+} + 2BH_4^- + 2H_2O \rightarrow Fe(0)(s) + 2B(OH)_3 + 7H_2(g) \qquad (1.15)$$

The sodium borohydride can be added in high excess with respect of the ferric or ferrous ion in order to achieve a rapid and uniform growth of iron crystals. The synthesis at lower concentrations is also satisfactory (Ponder, 2003). Although it is a physically simple process, reduction of aqueous metal salts through borohydrides is a complex reaction, sensitive to a variety of parameters, such as pH (which affects the particle size), and the concentration of the borohydride solution and its addition rate (which alter the composition of the reaction product). In some studies, it has been proved that the obtained particles were generally lower that $0.2\,\mu m$ (mostly between 1 and 100 nm) (Elliott and Zhang, 2001), with a specific area around $35\,m^2\,g^{-1}$ (Shen *et al.*, 1993) or close to $60\,m^2\,g^{-1}$ (Morgada *et al.*, 2009), whereas the commercial iron only attains $0.9\,m^2\,g^{-1}$ (Nowack and Bucheli, 2007). Some precautions must be taken during the process of synthesis of the nanoparticles to avoid the oxidation of iron due to the presence of oxygen, water and salts formed in the process of reduction (García *et al.*, 2008). The method produces highly reactive nZVI; however, the nanoparticles are often highly polydispersed, ranging over tens to hundreds of nanometers in size and thus significantly prone to agglomeration (Nurmi *et al.*, 2005; Scott *et al.*, 2010; Sun *et al.*, 2006). Expensive reagents and production of large volumes of hydrogen gas also preclude its industrial application (Hoch *et al.*, 2008). Scanning electron microscopy images of Fe(s) particles obtained by reduction with sodium borohydride shown particle size between 20 and 120 nm, with a medium size of 77 nm for those obtained from Fe(III) salts and of 87 nm for those obtained from Fe(II) salts (Elliott and Zhang, 2001).

The main obstacle in the preparation of nanoparticles is their tendency to aggregate and form particles of higher size, reducing their high surface energy. To avoid this, they can be prepared in the presence of a surfactant forming a microemulsion keeping the particles separated. The microemulsions form reverse micelles and are of especial interest because they can be introduced into a variety of reagents in the aqueous domains of nanometric size to produce reactions confined in the reverse micelles, achieving materials of controlled size and shape (Egorova and Revina, 2000). In these systems, the aqueous phase is dispersed in form of microdrops, in whose nuclei

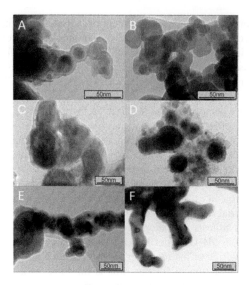

Figure 1.5.   Transmission electron microscopy (TEM) images of different iron nanoparticles manufactured or purchased. (A) nZVI synthesized by reduction of aqueous Fe(II) with sodium borohydride (Wang and Zhang, 1997); (B) nanoscale magnetite, purchased from Sigma–Aldrich; (C) NANOFER STAR, purchased from NANO IRON, s.r.o.; (D) nZVI synthesized by the carbothermal reduction of aqueous Fe(II); (E) nZVI synthesized by the reduction of aqueous Fe(II) using green tea polyphenols; and (F) nZVI synthesized by the reduction of aqueous Fe(II) using sodium borohydride then annealed under vacuum (at least 10^{-6} mbar) at 500°C for 24 h (Crane and Scott, 2012).

takes place the process of aggregation, and the particles in the growing phase are surrounded by the surfactant molecules (Colvin, 2003; Robinson *et al.*, 1989). In this way, the formation of particles and their dimensions can be controlled in a simple way and without high costs (Colvin, 2003). Another way to get lower size particles is by the use of ultrasound in the process of formation of the nanoparticles, because this helps to their separation.

In recent years, there has been much investigation to produce nZVI at accessible prices whilst maintaining reactivity and/or functionality. A method using hydrogen as a reducing agent for iron oxide nanoparticles at 350–600°C was patented in 2006 (Uegami *et al.*, 2006). Another material for large-scale field deployment of nZVI was produced by a company by the mechanical attrition of macroscale Fe(0) in planetary ball mill systems (USEPA, 2010), although the method is highly energy intensive and the particles exhibit a very high surface energy and are thus prone to aggregation.

Transmission electron microscopy images of different iron nanoparticles manufactured or purchased can be seen in Figure 1.5 (Crane and Scott, 2012).

1.6.3   *Bimetallic nZVI particles*

The chemical reactivity of nZVI can be improved by alloying them with a second metal such as Pd, Pt, Ag, Ni or Cu. Some studies have been made recently in the preparation of Fe/Pt (Zhang *et al.*, 1998), Fe/Ag (Xu and Zhang, 2000), Fe/Pd (Lien and Zhang, 2007), and Fe/Ni (Barnes *et al.*, 2010). In these electrochemical couples, Fe(0) behaves as an anode, becoming sacrificially oxidized to galvanically protect the second (nobler) metal. Chemical reduction of sorbed contaminants at the bimetallic nZVI surface is believed to occur through either direct electron transfer with the noble metal or through reaction with hydrogen produced by Fe(0) oxidation. For the clean-up of chlorinated organic contaminants such as TCE or PCP, hydrogen is observed to be the predominant driver for degradation, by breaking C–Cl bonds and swapping itself for chlorine, which is liberated as a gas. An advantageous consequence is that dichloroethelyenes (DCE) and vinyl chloride (VC), which are generated by TCE breakdown, are rapidly hydrogenated at the particle surfaces and do not accumulate in the reaction (Schrick *et al.*, 2002). In the US, approximately 50% of all nZVI remediation projects use standard nZVI and 40% use bimetallic nZVI; in Europe, no field application using bimetallic particles is yet to take place (Müller *et al.*, 2012).

Bimetallic nZVI reactivity depends on a range of factors, including nanoparticle size, physicochemical properties, and choice and quantity of the nobler metal. Minimal improvement with respect to their monometallic counterpart has been documented in some studies have (Dickinson and Scott, 2010) whilst others report an enhancement by several orders of magnitude (Tee *et al.*, 2009). Bimetallic nZVI could be used in preference to monometallic nZVI if they offer significantly improved performance at a competitive price (Pt, Pd and Ag are too costly), and if they guarantee that they do not add ecotoxicity by the inclusion in the treated systems. Because of corrosion, longevity of the materials can be poor. Consequently, bimetallic nZVI may be best suited for remediation applications where only short migration times to the contaminant plume are required.

1.6.4 *Stability of metal nanoparticles*

One of the most frequently questions in technology of nanoparticles is their long-term performance and reactivity. Due to their reduced size and high surface area, nanoparticles react very rapidly with a large variety of oxidants in groundwater, including dissolved oxygen, natural organic matter and water. It is speculated that the nanoparticles can have a limited lifetime in the subterranean environment (Chatterjee, 2008; Kanel *et al.*, 2006). In experiments carried out with iron nanoparticles, significant surface oxidation after reduced periods were observed, changing their color from dark to pale brown. Some experiments have demonstrated that the reactivity decreases substantially after air exposure for some days.

However, a change of color much more reduced has been observed in bimetallic nanoparticles of iron and palladium. ZVI nanoparticles underwent surface oxidation within a few hours (black to reddish-brown), whereas Pd modified iron particles did not undergo an observable color change in air suggesting stability (Zhang *et al.*, 1998). Correspondingly, ZVI particles lose reactivity within a few days, while Fe/Pd particles remain active for at least two weeks (Lien and Zhang, 2001). These bimetallic particles have demonstrated enough stability under environmental conditions, and it is expected that they can keep their reactivity for extended periods in the subterranean environment (USEPA, 2007).

Studies performed about the content of water of the iron particles suggest that there is a limit of life for the nanoparticles. Due to the corrosion produced by the medium, the estimated lifetime for metal zerovalent nanoparticles will be more limited than that of the iron particles of lower surface area. Iron nanoparticles contain considerably higher amount of water, linked physically and chemically, than iron particles of larger size. Whereas the nanoparticles do not lose their reducing power for a period of one year or more, a long-term exposure to air can finally produce dehydration forming less porous and less reactive surface oxides (SenGupta *et al.*, 2003).

Although it is generally recognized that iron nanoparticles are powerful in removal of pollutants, the colloidal chemistry of these particles makes them prone to agglomeration. For this reason, much work is being done on the immobilization of these particles to avoid agglomeration. A disadvantage of nZVI is the agglomeration of particles to each other and fast attachment of agglomerates to the soil surface. Agglomeration may be caused by groundwater conditions (pH, ionic strength), surface properties of the particles, the age of the materials, or shipping conditions (Saleh *et al.*, 2008; Zhang, 2003). Modifications to nanoscale iron particles have been made by using polymers or surfactants to enhance their mobility, reactivity, or stability. Surface modifiers increase the surface charge of the nanoparticles thereby providing electrostatic stabilization. They can also create a surface brush layer that engenders long-range strong steric repulsion forces, usually insensitive to high ionic strengths for which double layer repulsions would be greatly shielded. Examples of iron nanoparticles stabilization follow:

- Coatings such as polyelectrolytes or triblock polymers can be added in the suspension (He et al., 2009; Hydutsky et al., 2007; Lin et al., 2010; Saleh et al., 2007; 2008; Wang and Roman, 2011; Zhang, 2003) to stabilize the iron nanoparticles and improve their mobility.
- The nanoparticles are encased in emulsified vegetable oil droplets (EZVI) (Quinn et al., 2005) to improve stability and to improve reactivity by improving contact with the contaminant.
- Multi-functional nanocomposites (MFNC) are developed by incorporating nZVI into porous submicron particles (nanocomposites) of silica (Zhan et al., 2008; Zheng et al., 2008) or carbon (Zhan et al., 2011), to prevent agglomeration and couple iron reactivity with high silica/carbon adsorptivity.
- Guar gum is used as the stabilizing agent of aqueous suspension to reduce the attachment efficiency of NP in soil grains under varying conditions (Tiraferri and Sethi, 2009).
- Some nanoscale materials are made with catalysts (bimetallic nanoparticles, BNP) that enhance the intrinsic reactivity of the surface sites (Tratnyek and Johnson, 2006). In bench-scale tests, BNPs of iron combined with Pd achieved contaminant degradation two orders of magnitude greater than microscale iron particles alone: these particles were 99.9% iron and less than 0.1% Pd (Zhang et al., 2011). BNPs are generally incorporated into a slurry which may be stabilized by a combination of the aforementioned procedures (Schrick et al., 2004; Zhan et al., 2009), and can be injected by gravity or pressure feed.

Research indicates that nanoscale materials such as nZVI, BNPs, EZVI, MFNC may chemically reduce effectively PCE, TCE, cis-1,2-dichloroethylene, vinyl chloride, and 1-1-1-tetrachloroethane, polychlorinated biphenyls, halogenated aromatics, nitro-aromatics, heavy metals such as, lead and chromium, and arsenic, nitrate, perchlorate, sulfate, and cyanide (Bennett et al., 2010; Cundy et al., 2008; Kanel et al., 2005; Ponder et al., 2000; Zhang, 2003).

1.6.5 Other zerovalent nanoparticles used in soil and groundwater remediation

While iron is the most extensively examined metal for reductive processes like dechlorination, other metals such as Cu can also degrade halogenated hydrocarbons (Liou et al., 2007). Although nanosized and microsized ZVI are capable of degrading a wide array of highly chlorinated contaminants, e.g., trichloroethylene, the reactivity of ZVI towards as for example 1,2-dichloroethane is very low, but Cu zerovalent nanoparticles are rather effective (Huang et al., 2011). Moreover, unlike nZVI, Cu nanoparticles exhibit relative stability in water, avoiding unwanted reactions with surrounding media.

Various types of metal-oxides nanopowders (Al_2O_3, TiO_2, Fe_3O_4, MgO, ZrO_2, etc.) have been tested as adsorbents of arsenate ions from groundwater (Hristowski et al., 2007). Nanocrystalline metal salts have recently been tested to immobilize heavy metals in soils, sediments, and groundwater; for example, iron phosphate nanoparticles was tested for Pb (Liu and Zhao, 2007), iron sulfide nanoparticles for Hg (Xiong et al., 2009), and nanohydroxyapatite for Cd and Pb (Zhang et al., 2010). Nanoparticles of iron (oxyhydr)-oxides of controlled shape and size (goethite nanorods, hematite nanocubes, magnetite nanoparticles) were obtained from amorphous hydrous ferric- or ferrous-oxide by using a cost-effective hydrothermal method. These materials demonstrated to have 6–50 times higher adsorption capacity for As(III) compared to microsized particles (Zhao et al., 2011). The use of clays and iron-oxide minerals as nanocatalysts of Fenton-like reactions is a promising alternative for the decontamination of soils, groundwater, and sediments (Garrido-Ramirez et al., 2010; Zelmanov and Semiat, 2008). Recent studies suggest that nanoparticles of calcium peroxide (CaO_2) are a more effective source of hydrogen peroxide for in-situ chemical oxidation of organic pollutants dissolved in groundwater (Khodaveisi et al., 2011). Nanoparticulate magnetite is a potentially important reductant for environmental contaminants, including several halogenated organics and heavy metals (Gorski et al., 2010; Lee and Batchelor, 2002; McCormick *et al.*, 2002; Scott *et al.*, 2005). Magnetite shows reactivity toward CCl$_4$. Particle diameter plays an important role, with nominal 9 nm magnetite suspensions exhibiting greater reactivity than 80 nm magnetite suspensions (Vikesland *et al.*, 2007).

1.6.6 *Field application of nZVI injection in subsurface*

Nanoparticles are typically injected as slurries (nanofluids) directly into the subsurface environment to remediate contaminated groundwater plumes or contaminant source zones and may be suspended in the injected fluid to prevent particle agglomeration and enhance reactivity and mobility. Toward this direction, a variety of coatings (e.g., polyelectrolytes, surfactants, polymers, etc.) and supports (e.g., carbon, silica) have been used to stabilize nanofluids by increasing their resistance to particle aggregation and facilitating their delivery to target pollutants. According to the filtration theory, the mobility of colloidal particles in soils is governed by particle properties (e.g., size distribution, shape, electric charge, concentration) and hydraulic and physicochemical properties of soil and its environment (e.g., matrix chemical composition, surface zeta potential, temperature, pore structure, pH, ionic strength, groundwater composition, etc.) (Uyusur *et al.*, 2010). The foregoing parameters are of key importance for the Brownian diffusion of particles, their sedimentation and interception, and subsequently for the fate of nanoparticles during their transport through soil pores. Therefore, there is a need to establish nanotechnology synthesis routes that will control the mobility/reactivity/stability/toxicity of nanoparticles by adapting the nanofluids composition to the properties of soil, groundwater, and pollutants. No systematic study has ever been done to specify the most suitable agents that ensure the stabilization of adsorptive/reactive nanomaterials and their successful delivery to the target pollutants within contaminated soils and groundwater.

Karn *et al.* (2009) have compiled a comprehensive overview of some of the sites treated with nZVI. Most of these sites are located in the USA (Benett *et al.*, 2010; He *et al.*, 2009; Quinn *et al.*, 2005), and details can be viewed at the website of the Project on Emerging Nanotechnologies (Kuiken, 2010). In Europe, only a small number of pilot studies and a few full-scale remediation projects have been conducted (e.g., in the Czech Republic, Italy, and Germany). Before carrying out a full-scale application of nZVI, precise site investigations and pilot tests are needed, including the site hydrogeology as well as the geochemistry (Karn *et al.*, 2009). The hydrogeology influences the leachability of the particles while the geochemistry indicates potential substances that nZVI could react with other than the target compounds and thus determines the longevity of the reactive particles. Pilot tests are conducted to provide information on the amount of nZVI needed and possible unanticipated challenges. Information about the pilot-tests performed in Europe is shown in Table 1.4.

In Europe, full-scale demonstration of the *in-situ* groundwater remediation by injecting nZVI has been done on three sites (Müller and Nowack, 2012) contaminated by chlorinated solvents (PCE, TCE, DCE): (i) Bornheim (Rhein-Sieg-Kreis, Germany); (ii) Horice (Czech Republic); (iii) Pisecna site (Czech Republic). High pollutant remediation efficiencies (80–90%) were confirmed at a total cost (including treatment and monitoring) of € 300,000–360,000 (Müller and Nowack, 2010; Müller *et al.*, 2012).

All field applications carried out in Europe targeted groundwater only, while in the USA, about half of the site remediation targeted groundwater alone. About one fifth treated groundwater and soil simultaneously and a small number of site remediation treated sands, clayey silts, or soils (Karn *et al.*, 2009). In Europe, nZVI was in most cases injected into high permeability aquifers (more than one third of sites), 25% targeted fractured bedrock and only a few pilot projects were carried out in low permeability aquifers, unconsolidated sediments, or sandy gravel. For another 25% of the projects, the structure of the subsurface was not reported. Generally, it is agreed that remediation with nZVI in dense geological formations is less efficient and that unsaturated media are more difficult to treat. However, in these cases, hydraulic conductivity can be increased by fracturing and unsaturated zones can be flooded before or during the treatment. The range of possible applications of nZVI is wide as it can not only effectively degrade organic

Table 1.4. Pilot tests with nZVI in Europe (Müller et al., 2012).

Site	Date	Pollutant	Quantity of nZVI	[nZVI] (slurry)	Particle type	Injection technique	Media
Uzin, CZ	2009	Cl-ethenes	150 kg	$1-5\,g\,L^{-1}$	Nanofer	Infiltration drain	Low permeable aquifer
Rozmital, CZ	2007–2009	PCB	150 kg	$1-5\,g\,L^{-1}$	RNIP, Nanofer	Infiltration wells	Fractured bedrock
Spolchemie, CZ	2004, 2009	Cl-ethenes	20 kg	$1-10\,g\,L^{-1}$	Fe(B), Nanofer	Infiltration wells	Porous aquifer
Uhersky Brod, CZ	2008	Cl-ethenes	50 kg	$1-5\,g\,L^{-1}$	Nanofer	Infiltration wells	Porous aquifer
Hluk, CZ	2007, 2008	Cl-ethenes	150 kg	$1-5\,g\,L^{-1}$	RNIP, Nanofer	Infiltration wells	PRB filter
Hannover, DE	2007	CHC, BTEX, HC	1 kg	n.a.	n.a.	Infiltration wells	Unspecified
Asperg, DE	2006	Cl-ethenes	44 kg	$30\,g\,L^{-1}$	RNIP	Sleeve pipe	Fractured rock
Gaggenau, DE	2006	PCE	47 kg	$20\,g\,L^{-1}$	RNIP	Sleeve pipe	Porous aquifer
Permon, CZ	2006	Cr(VI)	7 kg	$1-5\,g\,L^{-1}$	RNIP	Infiltration wells	Fractured bedrock
Kurivody, CZ	2005, 2006	Cl-ethenes	50 kg	$1-10\,g\,L^{-1}$	Fe(B), RNIP	Infiltration wells	Fractured bedrock
Biella, IT	2005	TCE, DCE	10 kg	$1-10\,g\,L^{-1}$	nZVI	Gravity infiltration	Porous aquifer
Piestany, CZ	2005	Cl-ethenes	20 kg	$1-5\,g\,L^{-1}$	Fe(B)	Infiltration wells	High permeable aquifer
Schonebeck, DE	2005	Vinyl chloride	70 kg	$15\,g\,L^{-1}$	RNIP	Push infiltration	Porous aquifer
Thuringia, DE	2006	CAH, Ni, Cr(VI), nitrate	120 kg	$10\,g\,L^{-1}$	nZVI	Injection wells	Porous aquifer
Brownfield, SK	n.a.	TCE, DCE	n.a	n.a.	n.a.	n.a.	Unconsolidated sediments

n.a.: not annotated.

contaminants but also immobilize inorganic anions such as arsenic or chromium and can even be used to recover/remove dissolved metals from solution (Müller and Nowack, 2010). nZVI has been found to be effective also against PCB and organochlorine pesticides (Zhang, 2003). In Europe, most remedial actions with nZVI (70%) addressed chlorinated ethenes (PCE, TCE, DCE) or other chlorinated hydrocarbons (e.g., PCB). A few pilot remediation tests (20%) of other carbonaceous materials (BTEX, HC, VC) were additionally carried out. Ten percent of the remediation projects involved the immobilization of metals (Cr, Ni) and one pilot application also targeted nitrate. For 15 field-scale applications in the USA, nZVI was in most cases used to treat a source zone of TCE and daughter products, and some of the sites were contaminated with Cr(VI) (USEPA, 2005). It has also been promoted the treatment of source zones contaminated by dense non-aqueous phase liquids (DNAPL) especially chlorinated alkenes such as PCE and TCE (US Navy, 2010).

1.6.7 *Summary of advantages of the use of metal nanoparticles*

The advantages of nZVI may be summarized as follows (Müller and Nowack, 2010):

- Fast reaction: (i) short treatment time; (ii) less cost; (iii) less exposure for workers, fauna and flora.
- Complete reduction pathway to non-toxic byproducts: (i) less exposure for workers, fauna and flora.
- *In-situ* treatment: (i) less equipment and aboveground structures required; (ii) less costs.

1.7 PROBLEMS TO BE SOLVED IN THE TECHNOLOGY OF PERMEABLE REACTIVE BARRIERS WITH ZVI

Implementation of the technology of reactive permeable barriers with metallic iron faces still several challenges: (i) the production and accumulation of by-products generated by the chemical reactions involved in the elimination of contaminants due to the low reactivity of iron for the pollutants; (ii) the decline of the reactivity of iron with time, probably due to the formation of passive layers or to the precipitation of metallic hydroxides and carbonates, and (iii) engineering problems in the construction of metal barriers in deep aquifers.

A way to overcome some of these problems is to increase the reactivity of the active material of the barrier, thus increasing the rate of the reaction and retarding the obstruction of the pores. The surface area of the solid iron has a direct influence on the number of active surface sites faced to the contamination plume. By reducing the size of the iron particles, the surface area increases and consequently the rate of reaction. Increasing the specific surface area should also involve an increase in the fraction of iron atoms present in the surface of the particle, thus creating a more reducing ability per unit of mass. This could allow the use of smaller quantities of iron in the treatment of the contamination plume. The ability to reduce the volume of excavation by using barriers of less thickness and lower volumes of excavation is an important factor, because the excavation results in a greater economic cost.

As mentioned before, several laboratory batch studies have shown that supported iron nanoparticles are better than the commonly used commercial iron to eliminate various contaminants, although studies in real long-lasting conditions are still necessary. In terms of molar ratio, iron nanoparticles supported on polymer resins reduce the pollutant between 20 and 30 times more than commercial iron samples. Tests carried out for 60 days show that 90% of the reduction occurs during the first 48 hours, for supported or unsupported nanoparticles and also for the conventional granular size (Environmental KTN, 2008). The support itself disperses the iron particles, thereby increasing the total specific surface area, and provides a higher hydraulic conductivity avoiding the agglomeration of nanoparticles. Reduction by borohydride produces similar metallic iron nanoparticles in terms of size and surface area, regardless the material of the support used.

However, an important aspect is the added volume necessary to include the support material in the medium. Resins contain only about one quarter of the total amount of iron per gram in comparison with commercial iron, and the presence of the support increases the global mass by a factor of five. Even so, studies have shown that equivalent weights of nanoparticles supported on resins may reduce 20 times more Cr(VI), based on the iron present, than particles of granular iron, and this disparity increases when the concentration of the pollutant reduces.

However, it is difficult to determine the rate of corrosion of the iron particles, once various experiments have shown that both the support material and the method of preparation cause a significant effect on the electrochemical reaction of the composite materials (Henderson and Demond, 2007). Finally, the hydraulic conductivity is an important aspect for permeable barriers. Groundwater flow is highly sensitive to changes in permeability, and an installed barrier becomes useless if the hydraulic conductivity of the barrier is sufficiently different from the surrounding environment as to redirect the flow of pollutants to the outside of the barrier. It is expected that the behavior of supported nanoparticles be similar to the particles of granular iron, which has not been considered an important parameter in existing barriers.

1.8 ELECTROKINETICS

EK is based on the application of low-density electric currents between electrodes placed in the soil originating the migration of ions towards the corresponding electrode. In addition, several combinations of these technologies have been applied. The term electrokinetic remediation comprises, in fact, a number of different technologies. We consider essentially two major variants.

The first is based on the removal of ions (and/or polar organic compounds) using electromigration; in this method, the ions are mobilized under the direct action of an electric field. This approach was developed between 1958 and 1981 and has been marketed since 1988. The objective of the technology is to promote electromigration with external recirculation of the electrolyte pumped from compartments involving the electrodes created by physical walls permeable to ions but able to contain the anolyte and the catholyte. The second variant, initiated by Casagrande in 1947 and later developed by Honig in 1987 (Lageman *et al.*, 2005), favors the electroosmosis and is based on the movement of water through the electric double layer formed in the porous medium. Pollutants are transported through the water layer moving towards the cathode or in the direction of an adsorbent medium placed in its path, without recirculation of electrolytes. This approach was initiated at the Massachusetts Institute of Technology (MIT) in 1989. Of these two approaches, electromigration is gradually displacing the electroosmotic approach.

When electrodes submitted to a potential are submerged into an aqueous solution in a humid soil, the first phenomenon that occurs is the electrolysis of water. The solution becomes acidic at the anode due to hydrogen ion production and release of oxygen, and alkaline at the cathode as the consequence of the release of hydroxide anions and hydrogen (Fig. 1.6). At the anode, pH may fall to values below 2, while pH of the cathode can exceed 12, depending on the intensity of the applied current. At the cathode:

$$H_2O + e^- \rightarrow \tfrac{1}{2} H_2 + OH^- \tag{1.16}$$

At the anode:

$$H_2O \rightarrow 2H^+ + \tfrac{1}{2} O_2 + 2e^- \tag{1.17}$$

Most soils are conductive due to the presence of ions dissolved in the soil water, such as calcium, sodium, potassium, carbonates, fatty acids, nitrates, phosphates, sulfates and chlorides. A moisture content of 5% is sufficient to allow movement of these ions. Then, under the action of electric current, an acid front moves from the anode to the cathode by migration and advection causing desorption of contaminants from the soil. Since the H^+ ion has a much higher mobility than other ions, it will carry a disproportionately higher fraction of total current. The process

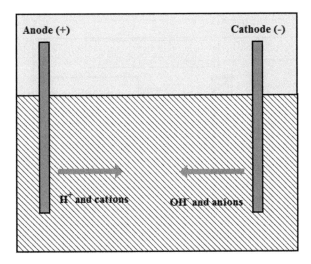

Figure 1.6. Movement of ions in EK.

results in a temporary acidification of the soil zone that is treated. However, it is not possible to theoretically predict the time required to establish equilibrium. The mobility of the hydronium cation is about two times higher than that of the hydroxide ion, and significantly higher than that of metal ions. However, in the soil, the hydronium cation is rapidly sorbed at the available ion exchange sites where the metal ions are in turn released to the electrolyte. At the same time, there is a similar movement, but slower and in the opposite direction, of hydroxide anions from the cathode to the anode, desorbing anions attached to the soil particles. The encounter of metal cations with hydroxide anions (since they are moving in opposite directions) may cause precipitation. In practice, however, this is avoided by making the catholyte slightly acidic.

The main factors to consider in designing a facility for As removal by electrokinetics are as follows:

- Creation of anode and cathode compartments through ion permeable housing (anionic or cationic membranes) with a diameter of 100–120 mm. These are placed in the contaminated medium and connected to a centralized management system of electrolytes. Each housing contains an electrode. The system is built with alternate rows of anodes and cathodes. The electrolytes are circulated in a closed circuit between the electrode sheaths and the management system of electrolytes (MSE). Through these electrolytes, the pH is maintained at a specified value.
- A potential difference is then applied to the electrodes. The water undergoes electrolysis at the electrodes, forming H^+ and O_2, which is generated at the anode, and OH^- and H_2, generated at the cathode. These ions migrate from the electrodes housing to the soil creating a considerable (but temporary) variation of the pH. This allows the desorption of the ionic contaminants. As such, it is not necessary to inject acid to the subsurface.
- Once desorbed, the contaminant ions migrate to the respective electrodes under the action of the applied voltage gradient (electromigration, Fig. 1.7). The anions migrate toward the anode and the cations to the cathodes. They flow through the housings of the electrodes and enter the electrolytes circulation system.
- The critical point of the management system is the careful control of pH and other conditions of the electrolyte in the electrodes housing;
- The contaminants are recovered from the solution obtained from the circulation of the electrolytes.

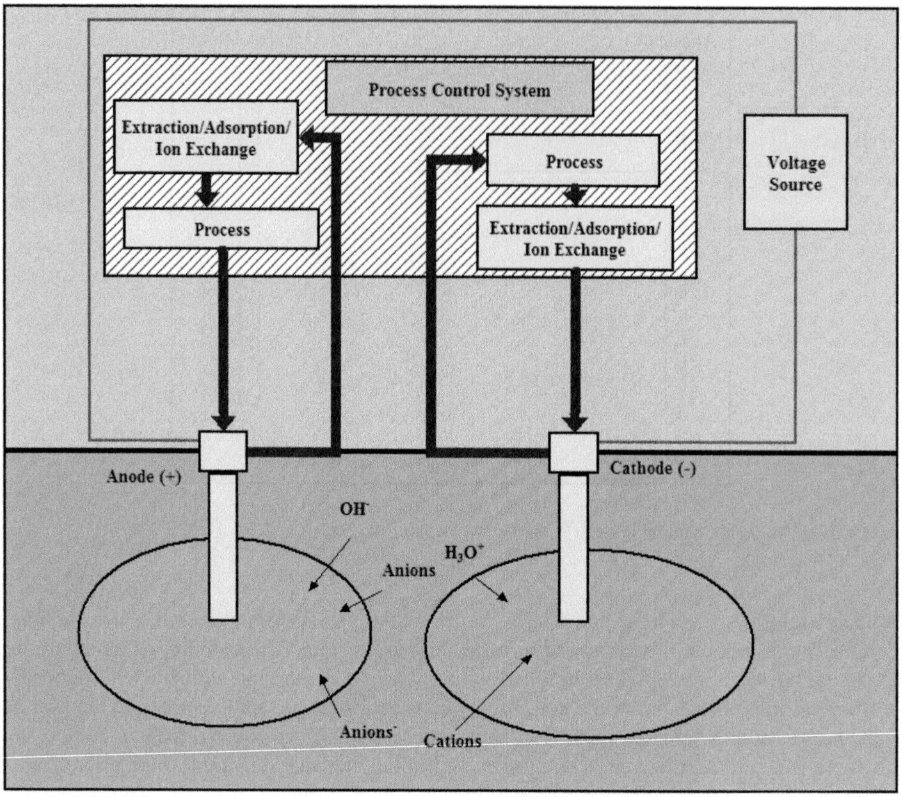

Figure 1.7.   Diagram of EK based on electromigration (adapted from USEPA, 1997).

When the conductivity of the electrolyte reaches about $20 \, \text{S} \, \text{m}^{-1}$, the electrolytes are pumped to a treatment plant. Depending on the type of contaminants, they can be treated by:

- NaOH to precipitate metal hydroxides, which are then removed by a press filter. The amount of cake produced depends not only on the concentration of heavy metals but also on the concentrations of iron, alkaline and alkaline earth metal ions, carbonates, hydrogencarbonates, magnesium, etc. As a rule, between 0.05 and 0.1% of the volume of the treated soil is collected as a residue;
- Ion exchange;
- Another adsorbent (in the case of As, activated alumina, iron oxides and hydroxides, etc.).

In alkaline medium, As species are poorly adsorbed, although As(V) is more easily adsorbed than As(III). The alkaline conditions favor the electromigration of As, although this is very slow (Virkutyte *et al.*, 2002). To improve the process of electromigration, sodium hypochlorite is added to the cathode compartment. In experiments made by Hécho *et al.* (1998), hypochlorite was added directly to the cathode compartment and As began to be collected in the anolyte after 15 days. After the remediation, the rate increased rapidly: after 20 days, 60% of As had been eliminated. However, following this, the elimination rate decreased again. Complete removal took 41 days, when the analysis showed that the residual concentration of As in the soil was less than 1 ppm.

Another alternative is the addition of complexing agents such as EDTA, which compete favorably with the soil for metal capture. Usually, a combination of three types of reagents is added: chelating agents, acidic solutions and cationic surfactants. The results presented by

Yuan and Chiang (2008) tested comparatively EK in the same soil under four different operating conditions: (i) groundwater only, and adding (ii) cetylpyridinium chloride, (iii) citric acid and (iv) EDTA. Systems with additives showed better performance than the groundwater alone. The best performance was observed in the EK-EDTA system with a gradient voltage of $3.3\,V\,m^{-1}$, and at the same time, there was a decrease in the electroosmotic permeability. The results indicated that the removal efficiency depends more on the mechanisms of electromigration rather than on the electroosmosis mechanisms. Indeed, the intense electroosmotic flow in the cathode caused a delay in the electromigration of As towards the anode. The amount of As collected in the anode compartment was 2.4 times higher than the amount collected in the cathodic compartment.

Leszczynska and Ahmad (2006) studied the electrokinetic removal of the pesticide known as copper and chromium arsenate (CCA). Studies were made with artificially contaminated kaolin using a DC power source. The authors compared the behavior of the system with and without the addition of reagents (NaOH and NaOCl). Doped systems behave more efficiently with removal of 74.4% and 78.1% for NaOH and NaOCl, respectively.

An As EK treatment plant was built in Loppersum, a small town in Northern Netherlands (Lageman, 2005). The soil at the site had a clayish nature and was contaminated with As with concentrations of around $400–500\,mg\,kg^{-1}$ to a maximum depth of $2\,m$. The source of the contamination was $Na_2HAsO_4 \cdot 7H_2O$, a compound used in wood preservation. There were two contaminated areas. Initially, a potential gradient of $40\,V\,m^{-1}$ was applied and then decreased to $20\,V\,m^{-1}$, with a current density of $4\,A\,m^{-2}$ (the total cross sectional area was $110\,m^2$). Ten monitoring sites were established. After 65 days, approximately 75% of the area had already a concentration below the target figure of $30\,mg\,kg^{-1}$. In short, they treated $250\,m^3$ of soil with an average As concentration of $115\,mg\,kg^{-1}$ and a maximum concentration of $500\,mg\,kg^{-1}$. The final average concentration was $10\,mg\,kg^{-1}$ and the maximum of $29\,mg\,kg^{-1}$. The energy consumption was $150\,kWh\,t^{-1}$. The duration of operation was 80 days of 18 hours, and the EK removed $38\,kg$ of As, while the remaining $14\,kg$ were removed by excavation.

1.9 *IN-SITU* CHEMICAL TREATMENT

In-situ chemical treatment consists of the direct injection of an oxidant in the subsoil (normally potassium permanganate or oxygen) promoting oxidation of As(III) to As(V), which then coprecipitates with iron oxides. All the set of *in-situ* chemical treatment technologies have in common the injection of a chemical reagent within an aquifer upstream of the contaminated site. This reacts with the contaminant and transforms it into a harmless form. Eventually, a closed loop can be used by pumping a certain volume of groundwater downstream the site and use it for reinjection. It is important to increase in the groundwater the velocity through the contaminated area by increasing the hydraulic gradient obtained by the actions of injection (water table elevation) and extraction (low water table). On the other hand, it is important to monitor the chemical transformation of the contaminant by chemical reaction within the treatment area.

In the case of As, the most efficient technology involves the introduction in the aquifer of two solutions: one of hydrogen peroxide and another one of ferric chloride. Hydrogen peroxide oxidizes As(III) to As(V) according to the following reaction:

$$HAsO_2 + 2H_2O \rightarrow H_3AsO_4 + 2H^+ + 2e^- \quad (E^0 = 0.56\,V) \tag{1.18}$$

Ferric chloride stabilizes As(V) by coprecipitation in the form of ferric arsenate and other insoluble precipitates. The precipitation reaction is:

$$Fe^{3+} + AsO_4^{3-} \rightarrow FeAsO_4 \quad (K_{ps} = 5.7 \times 10^{-21}) \tag{1.19}$$

Both solutions can be injected sequentially. As the oxidation reaction by the peroxide has a rather rapid kinetics, $FeCl_3$ can be injected shortly thereafter.

This technology was used in the remediation of a site in Tacoma (Ipsen *et al.*, 2005b), Washington, USA, where the groundwater contamination was caused by sodium arsenite. The solutions were injected using either direct-push injection (direct injection) with a spacing between the wells of 4.5 m, or by injection wells for the introduction of larger volumes. The design of the injection system is the essential criterion for the success of the technology.

Another alternative available technology is *in-situ* chemical bonding. The purpose of this technology (via injection of appropriate reagents in the subsoil) is to favor the formation of insoluble solid phases and drastically reduce their solubility. For example, soils with relatively soluble As compounds (as in the case of application of the herbicide As_2O_3) can be treated by introducing reagents, such as $FeSO_4$ and $KMnO_4$, which form insoluble As forms. These reagents drastically reduce the ability to leach As. Several studies examined the application of various solutions and concluded that the most efficient reagent was ferrous sulfate (Yang *et al.*, 2007). Seidel *et al.* (2005) studied the binding of As in the tailings through the formation of iron precipitates within a tailings dam. Iron precipitates were formed in situ by the aerobic treatment of ferrous sulfate solutions that were injected into the site.

1.10   COMBINATION OF ELECTROKINETICS AND PRB

Another approach that has been tested experimentally (Yuan and Chiang, 2007) is the combination of electrokinetics with a PRB. These researchers tested at laboratory scale the insertion of a PRB within a electrokinetic cell. The best results were obtained for a barrier built with FeOOH, using a gradient voltage of $2\,V\,cm^{-1}$. Electromigration predominates over the electroosmotic flow and As is removed by adsorption and precipitation over iron hydroxides, being found in a passivating layer that wraps the granules of hydroxides.

1.11   CONCLUDING REMARKS

In-situ technologies are suitable for groundwater treatment due to several advantages. They are environmental friendly and less expensive compared with conventional technologies (excavation, transport and *ex-situ* treatment of the contaminated soil). It can be applied in places that cannot be reached by excavation, rendering less polluted wastes.

However, some problems arise at present due to the novelty of these technologies. For example, handling, mixing and injection of the suspension are more expensive than handling of a solution, the injection covers a limited radius and more sophisticated injection equipment is needed, especially for nZVI.

However, the optimal method will come from the characteristics of the site. Generally, *in-situ* treatments are optimum where there is only one contaminant at low concentrations and where highly permeable aquifers have to be treated.

It is possible that *in-situ* treatments are not cost-effective for large dispersed plumes. There are also constraints coming for regulations and laws of environmental agencies that may differ from country to country or even from state to state.

Difficulty in getting approval for field tests by state agencies, to find funding for pilot projects, lack of information about *in-situ* remediation by consultants and potential clients, reluctant to use new technologies because of their inherent risk, lack of long term experience with the technology, etc. are other of the possible disadvantages.

In conclusion, much more research and development have to be devoted to improve the applicability of these *in-situ* techniques which appear to be rather useful to solve pollution problems hard to be solved by conventional techniques.

REFERENCES

Alvarez, P.: Chemistry and microbiology of permeable reactive barriers for in situ groundwater cleanup. *Crit. Rev. Microbiol.* 26 (2000), pp. 221–264.

Baker, M.J., Blowes, D.W. & Ptacek, C.J.: Laboratory development of permeable reactive mixtures for the removal of phosphorous from onsite wastewater disposal systems. *Environ. Sci. Technol.* 32 (1998), pp. 2308–2316.

Bang, S., Korfiatis, G.P. & Meng, X.: Removal of arsenic from water by zero-valent iron. *J. Hazard. Mater.* 121 (2005), pp. 61–67.

Barnes, R.J., Riba, O., Gardner, M.N., Scott, T.B., Jackman, S.A. & Thompson, I.P.: Optimization of nano-scale nickel/iron particles for the reduction of high concentration chlorinated aliphatic hydrocarbon solutions. *Chemosphere* 79 (2010), pp. 448–454.

Beak, D.G. & Wilkin, R.T.: Performance of a zerovalent iron reactive barrier for the treatment of arsenic in groundwater: Part 2. Geochemical modeling and solid phase studies. *J. Contam. Hydrol.* 106 (2009), pp. 15–28.

Benner, S.G., Blowes, D.W., Gould, W.D., Herbert, R.B. Jr. & Ptacek, C.J.: Geochemistry of a permeable reactive barrier for metals and acid mine drainage. *Environ. Sci. Technol.* 33 (1999), pp. 2793–2799.

Bennett, P., He, F., Zhao, D., Aiken, B. & Feldman, L.: In situ testing of metallic iron nanoparticle mobility and reactivity in a shallow granular aquifer. *J. Contam. Hydrol.* 116 (2010), pp. 35–46.

Bhumbla, D.K. & Keefer, R.F.: *Arsenic mobilization and bio-availability in soils. Arsenic in the environment. Part I: Cycling and characterization.* John Wiley & Sons, New York, NY, 1994, pp. 51–82.

Bianchi-Mosquera, G.C., Allen-King, R.M. & Mackay, D.M.: Enhanced degradation of dissolved benzene and toluene using a solid oxygen-releasing compound. *Groundwater Monit. Rem.* 14 (1994), pp. 120–128.

Blodau, C.: A review of acidity generation and consumption in acidic coal mine lakes and their watersheds. *Sci. Total Environ.* 369 (2006), pp. 307–332.

Blowes, D.W. & Ptacek, C.J.: *System for treating contaminated groundwater.* US Patent #5514279, 1996.

Blowes, D.W., Ptacek, C.J. & Jambor, J.L.: *In-situ* remediation of Cr(VI)-contaminated groundwater using permeable reactive walls: laboratory studies. *Environ. Sci. Technol.* 31 (1997), pp. 3348–3357.

Blowes, D.W., Ptacek, C.J., Benner, S.G., McRae, C.W.T., Bennet, T.A. & Puls, R.W.: Treatment of inorganic contaminants using permeable reactive barriers. *J. Contam. Hydrol.* 45 (2000), pp. 123–137.

Bolzicco, J., Carrera, J., Ayora, C., Ceron, J.C. & Fernández, I.: *Comportamiento y evolución de una barrera geoquímica experimental Río Agrio – Aznalcóllar – España.* CSIC, Barcelona, Spain, 2001.

Burghardt, D., Simon, E., Knöller, K. & Kassahun, A.: Immobilization of uranium and arsenic by injectible iron and hydrogen stimulated autotrophic sulphate reduction. *J. Contam. Hydrol.* 94 (2007), pp. 305–314.

Chatterjee, R.: The challenge of regulating nanomaterials. *Environ. Sci. Technol. 42 (*2008), pp. 339–343.

Chen, X., Wright, J.V., Conca, J.L. & Peurrung, L.M.: Effects of pH on heavy metal sorption on mineral apatite. *Environ. Sci. Technol.* 31 (1997), pp. 624–630.

CL:AIRE. Design, installation and performance assessment of a zero valent iron permeable reactive barrier in Monkstown Northern Ireland. *Technology Demonstration Report (TDP3)*, 2001. http://www.claire.co.uk/ (accessed May 2012).

Colvin, V.: The potential environmental impact of engineered nanoparticles. *Nature Biotechnol.* 21 (2003), pp. 1166–1170.

Crane, R.A. & Scott, T.B.: Nanoscale zero – valent iron: Future prospects for an emergingwater treatment technology. *J. Hazard. Mater.* 211–212 (2012), pp. 112–125.

Cundy, A.B., Hopkinson, L. & Whitby, R.L.D.: Use of iron-based technologies in contaminated land and groundwater remediation: A review. *Sci. Total Environ.* 400 (2008), pp. 42–51.

Deng, B. & Hu, S.: Reductive dechlorination of chlorinated solvents on zerovalent iron surfaces. In: J.A. Smith & S.E. Burns (eds): *Physicochemical groundwater remediation*, New York, NY, Kluwer Academic, 2001, pp. 139–159.

Dickinson, M. & Scott, T.M.: The application of zero-valent iron nanoparticles for the remediation of a uranium-contaminated waste effluent. *J. Hazard. Mater.* 178 (2010), pp. 171–179.

Diels, L. & Vanbroekhoven, K.: Remediation of metal and metalloid contaminated groundwater. In: M.D. Annable, M. Teodorescu, P. Hlavinek & L. Diels (eds): Proceedings of the NATO Advanced Research Workshop on Methods and Techniques for Cleaningup Contaminated Sites, Sinaia, Romania, 9–11 October 2006, *Series: NATO Science for Peace and Security Series C: Environmental Security*, 2008, VIII, Springer, Netherlands, pp. 1–23.

Diels, L., Bastiaens, L., O'Hannessin, S., Cortina, J.L., Alvarez, P.J., Ebert, M. & Schad, H.: Permeable reactive barriers: a multidisciplinary approach of a new sustainable groundwater technology. *Proc. ConSoils 2003*, 12–16 May, Ghent, Belgium, 2003, pp. 123–132.

Driehaus, W., Jekel, M. & Hildebrandt, U.: Granular ferric hydroxide – a new adsorbent for the removal of arsenic from natural water. *J. Water SRT-Aqua* 47 (1998), pp. 30–35.

Egorova, E.M. & Revina, A.A.: Synthesis of metallic nanoparticles in reverse micelles in the presence of quercetin., *Colloids and Surfaces A: Physicochemical and Engineering Aspects*, 168 (2000), pp. 87–96.

Elliott, D.W. & Zhang, W.: Field assessment of nanoscale bimetallic particles for groundwater treatment. *Environ. Sci. Technol.* 35 (2001), pp. 4922–4926.

Environmental KTN: Priority technology area 9: *In-situ land remediation. Environmental knowledge transfer network business case 9.* Oxford, United Kingdom, 2008. http://www.environmental-ktn.com (accessed May 2008).

Eweis, J.B., Ergas, S.J., Chang, D.P. & Schroeder, E.D.: *Bioremediation principles.* McGraw-Hill, Boston, MA, 1998.

Fryar, A.E. & Schwartz, F.W.: Modeling the removal of metals from ground water by a reactive barrier: experimental results, *Water Resour. Res.* 30 (1994), pp. 3455–3469.

García, E., Cortina, J.L., Farrán, A., de Pablo J. & Martí V.: Stabilization of zero valent nanoparticules onto macroporous polymeric sorbents for removal of anionic contaminants. In: M. Cox (ed): *Advances in ion exchange for industry and research.* SCI., London, UK, 2008, pp. 241–247.

Garrido-Ramirez, E.G., Theng, B.K.G. & Mora, M.L.: Clays and oxide minerals as catalysts and nanocatalysys in Fenton-like reactions – a review. *Appl. Clay Sci.* 47 (2010), pp. 182–192.

Gavaskar, A.R., Gupta, B.M., Janosy, R.J. & O'Sullivan, D.: *Permeable barriers for groundwater remediation – design, construction, and monitoring,* Batelle Press, Columbus, OH, 1998.

Gavaskar, A., Tatar, L. & Condit, W.: Cost and performance report: nanoscale zero-valent iron technologies for source remediation. *Contract report CR-05-007-ENV*, 93043-4370, Naval Facilities Engineering Command, Engineering Service Center, Port Hueneme, CA, 2005.

Gorski, C.A., Nurmi, J.T., Tratnyek, P.G., Hofstetter, T.B. & Scherer, M.M.: Redox behavior of magnetite: implications for contaminant reduction. *Environ. Sci. Technol.* 44 (2010), pp. 55–60.

Gotpagar, J., Grulke, E., Tsang, T. & Bhattacharyya, D.: Reductive dehalogenation of trichloroethylene using zero valent iron, *Environ. Progress* 16 (1997), pp. 137–143.

Gu, B., Watson, D.B., Phillips, D.H. & Liang, L.: Performance evaluation of a permeable iron reactive barrier used for treatment of radionuclides and other inorganic contaminants. *EOS Trans., Am. Geophys. Union, Fall Meet.* 80 (1999), F366.

Gubert, O., de Pablo, J., Cortina, J.L. & Ayora, C.: Chemical characterisation of natural organic substrates for biological mitigation of acid mine drainage. *Water Res.* 38 (2004), pp. 4186–4196.

Guillham, R.W. & O'Hannesin, S.F.: Metal-catalyzed abiotic degradation of halogenated organic coumpounds. *IAH Conference on Modern Trends in Hydrogeology*, 10–13 May 1992, Hamilton, Ontario, Canada, 1992, pp. 94–103.

Hammack, R.W., Edenborn, H.M. & Dvorak, D.H.: Treatment of water from an open-pit copper mine using biogenic sulfide and limestone: a feasibility study. *Water Res.* 28 (1994), pp. 2321–2329.

Harter, T.: *Groundwater quality and groundwater pollution.* Publ. 8084, University of California, Division of Agriculture and Natural Resources, Oakland, CA, 2003.

He, F., Zhang, M., Qian, T. & Zhao, D.: Transport of carboxymethyl cellulose stabilized iron nanoparticles in porous media: column experiments and modeling. *J. Colloid Interface Sci.* 334 (2009), pp. 96–102.

Hécho, L.I., Tellier, S. & Astruc, M.: Industrial site soils contaminated with arsenic or chromium: evaluation of the electrokinetic method. *Environ. Technol.* 19 (1998), pp. 1095–1102.

Henderson, A.D. & Demond, A.H.: Long-term performance of zero-valent iron permeable reactive barriers: a critical review. *Environ. Eng. Sci.* 24 (2007), pp. 401–423.

Herbert, R.B., Benner, S.G. & Blowes, D.W.: Solid phase iron sulfur geochemistry of a reactive barrier for treatment of mine drainage. *Appl. Geochem.* 15 (2000), pp. 1331–1343.

Hoch, L.B., Mack, E.J., Hydutsky, B.W., Hershman, J.M., Skluzacek, J.M. & Mallouk, T.E.: Carbothermal synthesis of carbon-supported nanoscale zero-valent iron particles for the remediation of hexavalent chromium. *Environ. Sci. Technol.* 42 (2008), pp. 2600–2605.

Hristovski, K., Baumgardner, A. & Westerhoff, P.: Selecting metal oxide nanomaterials for arsenic removal in fixed bed columns: from nanopowders to aggregated nanoparticle media. *J. Hazard. Mat.* 147(2007), pp. 265–274.

Huang, C.-C., Lo, S.-L., Tsai, S.-M. & Lien, H.-L.: Catalytic hydrodechlorination of 1,2-dichloroethane using copper nanoparticles under reduction conditions of sodium borohydride. *J. Environ. Monit.* 13 (2011), pp. 2406–2412.

Hydutsky, B.W., Mack, E.J., Bekerman, B.B., Skluzacek, J.M. & Mallouk, T.E.: Optimization of nano- and microiron transport through sand columns using polyelectrolyte mixtures. *Environ. Sci. Technol.* 41 (2007), pp. 6418–6424.

Ipsen, S.-O., Gerth, J. & Förstner, U.: Identifying and testing materials for arsenic removal by permeable reactive barriers, *Consoil* (2005), Proceedings of the 9th international FZK/TNO conference on Soil Water Systems. 3–7 October 2005, Bordeaux, France, 2005a, pp. 1815–1817.

Ipsen, E., Slater, J.T., Wolf, F. & Magee, B.: *In-situ* chemical remediation of arsenic in groundwater using ferric chloride, *Consoil* 2005, Proceedings of the 9th international FZK/TNO conference on Soil Water Systems, 3–7 October 2005, Bordeaux, France, 2005b, pp. 1516–1525.

Jang, M., Min, S-.H., Park, J.K & Tlachac, E.: Hydrous ferric oxide incorporated diatomite for remediation of arsenic contaminated groundwater. *Environ. Sci. Technol.* 41 (2007), pp. 3322–3328.

Johnson, T.L., Scherer, M.M. & Tratnyek, P.G.: Kinetics of halogenated organic compound degradation by iron metal. *Environ. Sci. Technol.* 30 (1996), pp. 2634–2640.

Joshi, A. & Chaudhuri, M.: Removal of arsenic from groundwater by iron oxide-coated sand. *J. Environ. Eng.* 122 (1996), pp. 769–772.

Kanel, S.R., Greneche, J.M. & Choi, H.: Arsenic(V) removal from groundwater using nanoscale zero-valent iron as a colloidal reactive barrier material. *Environ. Sci. Technol.* 40 (2006), pp. 2045–2050.

Kanel, S.R., Manning, B., Charlet, L. & Choi, H.: Removal of arsenic (III) from groundwater by nanoscale zero-valent iron. *Environ. Sci. Technol.* 39 (2005), pp. 1291–1298.

Kaplan, D.I., Cantrell, K.J., Wietsma, T.W. & Potter, M.A.: Retention of zero-valent iron colloids by sand columns: application to chemical barrier formation. *J. Environ. Contr.* 25 (1996), pp. 1086–1094.

Karn, B., Kuiken, T. & Otto, M.: Nanotechnology and in situ remediation: a review of the benefits and potential risks. *Environ. Health Perspectives* 117 (2009), pp. 1823–1831.

Khodaveisi, J., Banejad, H., Afkhami, A., Olyaie, E., Lashgari, S. & Dashti, R.: Synthesis of calcium peroxide nanoparticles as an innovative reagent for in situ chemical oxidation. *J. Hazard. Mat.* 192 (2011), pp. 1437–1440.

Köber, R., Giarolli, F. & Dahmke, A.: Development of up and downstream permeable reactive barrier systems for remediation of arsenic and VOC groundwater contaminations. *Sixth International Symposium & Exhibition on Environmental Contamination in Central & Eastern Europe and the Commonwealth of Independent States*, Prague, Czech Republic, 2003.

Köber, R., Daus, B., Ebert, M., Mattush, J., Welter, E. & Dahmke, A.: Compost-based permeable reactive barriers for the source treatment of arsenic contaminations in aquifers: column studies and solid-phase investigations. *Environ. Sci. Technol.* 39 (2005), pp. 7650–7655.

Kuiken, T.: Cleaning up contaminated waste sites: is nanotechnology the answer? *Nano Today* 5 (2010), pp. 6–8.

Lackovic, J.A., Nikolaidis, N.P. & Dobbs, G.M.: Inorganic arsenic removal by zero valentiron. *Environ. Eng. Sci.* 17 (1999), pp. 29–39.

Lageman, R. & Clarke, R.L., Pool, W.: Electro-reclamation, a versatile soil remediation solution. *Eng. Geol.* 77 (2005), pp. 191–201.

Lee, W. & Batchelor, B.: Abiotic reductive dechlorination of chlorinated ethylenes by iron-bearing soil minerals. 1. Pyrite and magnetite. *Environ. Sci. Technol.* 36 (2002), pp. 5147–5154.

Leszczynska, D. & Ahmad, H.: Toxic elements in soil and groundwater: short-time study on electrokinetic removal of arsenic in the presence of other ions. *Int. J. Environ. Res. Public Health* 3 (2006), pp. 196–201.

Lien, H.L. & Wilkin, R.T.: High-level arsenite removal from groundwater by zero-valent iron. *Chemosphere* 59 (2005), pp. 377–386.

Lien, H.-L. & Zhang, W.-X.: Nanoscale iron particles for complete reduction of chlorinated ethenes. *Colloids Surfaces* A: *Physicochemical and Engineering Aspects* 191 (2001), pp. 97–105.

Lien, H.L. & Zhang, W.X.: Nanoscale Pd/Fe bimetallic particles: catalytic effects of palladium on hydrodechlorination. *Appl. Catal.* B: *Environ.* 77 (2007), pp. 110–116.

Lin, Y.-H., Tseng, H.-H., Wey, M.-Y. & Lin, M.-D.: Characteristics of two types of stabilized nano zero – valent iron and transport in porous media. *Sci. Total Env.* 408 (2010), pp. 2260–2267.

Liou, Y.H., Lo,S.L. & Lin, C.J.: Size effect in reactivity of copper nanoparticles to carbon tetrachloride degradation. *Water Res.* 41 (2007), pp. 1705–1712.

Liu, R. & Zhao, D.: Reducing leachability and bioaccessibility of lead in soils using a new class of stabilized iron phosphate nanoparticles. *Water Res.* 41 (2007), pp. 2491–2502.

Lo, I.M.C., Surampalli, R.Y. & Lai, K.C.K.: Zero-valent iron reactive materials, for hazardous waste and inorganics removal. American Society of Civil Engineers (ASCE), Reston, VA, 2007.

Ludwig, R.D., Smyth, D.J.A., Blowes, D.W., Spink, L.E., Wilkin, R.T. Jewett, D.G. & Weisener, C.J.: Treatment of arsenic, heavy metals, and acidity using a mixed ZVI-compost PRB. *Environ. Sci. Technol.* 43 (2009), pp. 1970–1976.

Ma, Q.Y., Traina, S.J., Logan, T.J. & Ryan, J.A.: Effects of aqueous Al, Cd, Cu, Fe(II), Ni, and Zn on Pb immobilization by hydroxyapatite. *Environ. Sci. Technol.* 28 (1994), pp. 1219–1228.

Mackenzie, P.D., Sivavec, T.M. & Horney, D.P.: Mineral precipitation and porosity losses in iron treatment zones. *213 th Meeting, American Chemical Society*, San Francisco, CA. Preprint Extended Abstracts, Division of Environmental Chemistry, 37 (1997), pp. 154–157.

McCormick, M. L., Bouwer, E. J. & Adriaens, P.: Carbon tetrachloride transformation in a model iron-reducing culture: relative kinetics of biotic and abiotic reactions. *Environ. Sci. Technol.* 36 (2002), pp. 403–410.

McRae, C.W.T.: *Evaluation of reactive materials for in situ treatment of arsenic III, arsenic V and Selenium VI using permeable reactive barriers: laboratory study*. MSc Thesis, University of Waterloo, Waterloo, Ontario, Canada, 1999.

Morgada, M.E., Levy, I.K., Salomone, V., Farías, S.S., López, G. & Litter, M.I.: Arsenic (V) removal with nanoparticulate zerovalent iron: effect of UV light and humic acids. *Catal. Today* 143 (2009), pp. 261–268.

Morrison, S.J., Mushovic, P.S. & Niesen, P.L.: Early breakthrough of molybdenum and uranium in a permeable reactive barrier. *Environ. Sci. Technol.* 40 (2006), pp. 2018–2024.

Müller, C. & Nowack, B.: Nano-zero valent iron-THE solution for soil and groundwater remediation?, Report of the Observatory NANO, 2010.www.observatorynano.eu (accessed July 2012).

Müller, N.C., Braun, J., Bruns, J., Cernik, M., Rissing, P., Rickerby, D. & Nowack, B.: Application of nanoscale zero valent iron (NZVI) for groundwater remediation in Europe. *Environ. Sci. Pollut. Res.* 19 (2012), pp. 550–558.

Nikolaidis, N.P., Dobbs, G.M & Lackovic, J.A.: Arsenic removal by zerovalent iron: field, laboratory and modeling studies. *Water Res.* 37 (2003), pp. 1417–1425.

Nowack, B. & Bucheli, T.D.: Occurrence, behavior and effects of nanoparticles in the environment. *Environ. Pollut.* 150 (2007), pp. 5–22.

Nurmi, J.T., Tratnyek, P.G., Sarathy, V., Baer, D.R., Amonette, J.E., Pecher, K., Wang, C., Linehan, J.C., Matson, D.W., Penn, R.L. & Driessen, M.D.: Characterization and properties of metallic iron nanoparticles: spectroscopy, electrochemistry, and kinetics. *Environ. Sci. Technol.* 39 (2005), pp. 1221–1230.

O'Hannesin, S.F. & Gillham, R.W.: Long-term performance of an in situ iron wall for remediation VOCs. *Groundwater* 36 (1998), pp. 164–170.

Phillips, D.H., Watson, D.B., Roh, Y. & Gu, B.: Mineralogical characteristics and transformations during long-termoperation of a zerovalent iron reactive barrier. *J. Environ. Qual.* 32 (2003), pp. 2033–2045.

Ponder, S.: Surface chemistry and electrochemistry of supported zerovalent iron nanoparticles in the remediation of aqueous metal contaminants. *Chem. Mater.* 13 (2003), pp. 479–486.

Ponder, S.M., Darab, J.G. & Mallouk, T.E.: Remediation of Cr(VI) and Pb(II) aqueous solutions using supported, nanoscale zero-valent iron. *Environ. Sci. Technol.* 34 (2000), pp. 2564–2569.

Powell, R.M. & Puls, R.W.: Proton generation by dissolution of intrinsic or augmented aluminosilicate minerals for in situ contaminant remediation by zero valence-state iron. *Environ. Sci. Technol.* 31 (1997), pp. 2244–2251.

Powell, R.M., Puls, W.P., Hightower, S.K. & Sabatini, D.A.: Coupled iron corrosion and chromate reduction: mechanisms for subsurface remediation. *Environ. Sci. Technol.* 29 (1995), pp. 1913–1922.

Ptacek, C.J., Blowes, D.W., Robertson, W.D. & Baker, M.J.: Adsorption and mineralization of phosphate from septic system effluent in aquifer materials. *Wastewater Nutrient Removal Technologies and Onsite Management Districts,* Conference Proceedings, Waterloo, Ontario, Canada, June 6, 1994.

Puls, R.W., Paul, C.J. & Powell, R.M.: The application of in situ permeable reactive (zerovalent iron) barrier technology for the remediation of chromate-contaminated groundwater: a field test. *Appl. Geochem.* 14 (1999), pp. 989–1000.

Quinn, J., Geiger, C., Clausen, C., Brooks, K., Coon, C., O'Hara, S., Krug, T., Major, D., Yoon, W.-S., Gavaskar, A. & Holdworth, T.: Field demonstration of DNAPL dehalogenation using emulsified zero-valent iron. *Environ. Sci. Technol.* 39 (2005), pp. 1309–1318.

Robinson, B.H., Khan-Lodhi, A.N. & Towey, T.: Microparticle synthesis and characterization in reverse micelles. In: M. Pilen (ed): *Structure and reactivity in reverse micelles*. Elsevier, Amsterdam, The Netherlands, 1989, pp. 199–219.

Roh, Y., Lee, S.Y. & Elless, M.P.: Characterization of corrosion products in the permeable reactive barriers. *Environ. Geol.* 40 (2001), pp. 184–194.

Sacre, J.A.: Treatment walls: a status update. Ground-water Remediation Technologies Analysis Center, TP-97-02. Pittsburgh, 1997.http://www.gwratc.org (accessed June 2012).

Saleh, N., Sirk, K., Liu, Y., Phenrat, T., Dufour, B., Matyjaszewski, K. Tilton, R.D. & Lowry, G.V.: Surface modifications enhance nanoiron transport and NAPL targeting in saturated porous media. *Environ. Eng. Sci.* 24 (2007), pp. 45–57.

Saleh, N., Kim, H.-J., Phenrat, T., Matyjaszewski, K., Tilton, R.D. & Lowry, G.V.: Ionic strength and composition affect the mobility of surface-modified Fe0 nanoparticles in water-saturated sand columns. *Environ. Sci. Technol.* 42 (2008), pp. 3349–3355.

Scherer, M.M., Richter, S., Valentine, R.L. & Alvarez, P.J.J.: Chemistry and microbiology of permeable reactive barriers for in situ groundwater clean up. *Critical Rev. Microbiol.* 26 (2000), pp. 221–264.

Schneider, P., Neitzel, P.L., Osenbrück, K., Noubactep, Ch., Merkel, B. & Hurst, S.: *In-situ*-treatment of radioactive mine water using reactivematerials. *Acta Hydrochim. Hydrobiol.* 29 (2001), pp. 123–138.

Schrick, B., Blough, J.L., Jones, A.D. & Mallouk, T.E.: Hydrodechlorination of trichloroethylene to hydrocarbons using bimetallic nickel-iron nanoparticles. *Chem. Mater.* 14(2002), pp. 5140–5147.

Schrick, B., Hydutsky, B.W., Blough, J.L. & Mallouk, T.E.: Delivery vehicles for zerovalent metal nanoparticles in soil and groundwater. *Chem. Mater.* 16 (2004), pp. 2187–2193.

Scott, T.B., Allen, G.C., Heard, P.J. & Randell, M.G.: Reduction of U(VI) to U(IV) on the surface of magnetite, *Geochim. Cosmochim. Acta* 69 (2005), pp. 5639–5646.

Scott, T.B., Dickinson, M., Crane, R.A., Riba, O., Hughes, G.M. & Allen, G.C.: The effects of vacuum annealing on the structure and surface chemistry of iron nanoparticles. *J. Nanopart. Res.* 12 (2010), pp. 1765–1775.

Seidel, H., Görsch, K., Amstätter, K. & Mattusch, J.: Immobilization of arsenic in a tailings materialby ferrous iron treatment. *Water Res.* 39 (2005), pp. 4073–4082.

Sengupta, A.K.: Polymer supported inorganic nanoparticles: characterization and environmental applications. *Environ. Sci. Technol.* 54 (2003), pp. 167–180.

Sengupta, A.K., De Marco, M.J. & Greenleaf, J.E.: Arsenic removal using a polymeric/inorganic hybrid sorbent. *Water Res.* 37 (2003), pp. 164–176.

Shen, J., Li, Z., Yan, Q. & Chen, Y.: Reactions of bivalent metal ions with borohydride in aqueous solution for the preparation of ultrafine amorphous alloy particles. *J. Phys. Chem.* 97 (1993), pp. 8504–8511.

Silva, A., Freitas, O., Figueiredo, S., Vandervliet, B., Ferreira, A. & Fiúza, A.: Arsenic removal using synthetic adsorbents: kinetics, equilibrium and column study, *12th EuCheMS International Conference on Chemistry and the Environment*, Stockholm, Sweden, 2009.

Spira, Y., Henstock, J., Nathanail, P., Muller, D. & Edwards, D.A.: European approach to increase innovative soil and groundwater remediation technology applications. *Remediation* 16 (2006), pp. 81–96.

Su, C. & Puls, R.W.: In situ remediation of arsenic in simulated groundwater using zerovalent iron: laboratory column tests on combined effects of phosphate and silicate. *Environ. Sci. Technol.* 37 (2003), pp. 2582–2587.

Sun, Y.-P., Li, X.-Q., Cao, J., Zhang, W.-X. & Wang, H.P.: Characterization of zero-valent iron nanoparticles. *Adv. Colloid Interface Sci.* 120 (2006), pp. 47–56.

Tee, Y.-H., Bachas, L. & Bhattacharyya, D.: Degradation of tricholoroethylene by iron-based bimetallic nanoparticles. *J. Phys. Chem. C* 113 (2009), pp. 9454–9464.

Tiraferri, A. & Sethi, R.: Enhanced transport of zerovalent iron nanoparticles in saturated porous media by guar gum. *J. Nanopart. Res.* 11 (2009), pp. 635–645.

Tratnyek, P.G. & Johnson, R.L.: Nanotechnologies for environmental clean-up. *Nanotoday* 1 (2006), pp. 44–48.

Triszcz, J.M., Porta, A. & García Einschlag, F.S.: Effect of operating conditions on iron corrosion rates in zero-valent iron systems for arsenic removal. *Chem. Eng. J.* 150 (2009), pp. 431–439.

Tyrovola, K. & Nikolaidis, N.P.: Arsenic mobility and stabilization in topsoils. *Water Res.* 43 (2009), pp. 1589–1596.

Uegami, M., Kawano, J., Okita, T., Fujii, Y.,Okinaka, K., Kayuka K. & Yatagi, S.: *Iron particles for purifying contaminated soil or groundwater.* US Patent No. 7,022,256, April 4, 2006.

US Navy: *Nanoscale Zero Valent Iron*, 2010. https://portal.navfac.navy.mil/portal/page/portal/navfac/navfac_ww_pp/navfac_nfesc_pp/environmental/erb/nzvi (accessed May 2012).

USEPA: *Permeable reactive barrier technologies for contaminant remediation.* EPA 600/R-98/125. US. Environmental Protection Agency, Washington, DF, 1998.

USEPA 542-F-01-005: *A citizen's guide to permeable reactive barriers.* April 2001. http://www.clu-in.org/download/citizens/citprb.pdf (accessed May 2012).

USEPA: *USEPA Workshop on Nanotechnology for Site Remediation*, 2005. http://epa.gov/ncer/publications/workshop/pdf/10_20_05_nanosummary.pdf (accessed May 2005).

USEPA: *Nanotechnology white paper.* EPA 100/B-07/001. 20460: Science Policy Council, US Environmental Protection Agency, Washington, DF, 2007.

USEPA: 2010. http://www.clu-in.org/conf/tio/nano-iron 121410/ (accessed September 2011).

Uyusur, B., Darnault, C.J.G., Snee, P.T., Koken, E., Jacobson, A.R. & Wells, R.R.: Coupled effects of solution chemistry and hydrodynamics on the mobility and transport of quantum dot nanomaterials in the vadose zone. *J. Contam. Hydrol.* 118 (2010), pp. 184–198.

Vikesland, P.J., Heathcock, A.M., Rebodos, R.L. & Makus, K.E.: Particle size and aggregation effects on magnetite reactivity toward carbon tetrachloride. *Environ. Sci. Technol.* 41 (2007), pp. 5277–5283.

Virkutyte, J., Silanpää, M. & Latostenmaa, P.: Electrokinetic soil remediation – critical overview. *Sci. Total Environ.* 289 (2002), pp. 97–121.

Wang C.-B. & Zhang, W.-X.: Synthesizing nanoscale iron particles for rapid and completedechlorination of TCE and PCBs. *Environ. Sci. Technol.* 31 (1997), pp. 2154–2156.

Wang, H. & Roman, M.: Formation and properties of chitosan-cellulose nanocrystal polyelectrolyte-macroioncomplexes for drug delivery applications. *Biomacromolecules* 12 (2011), pp. 1585–1593.

Waybrant, K.R., Blowes, D.W. & Ptacek, C.J.: Selection of reactive mixtures for use in permeable reactive walls for treatment of mine drainage. *Environ. Sci. Technol.* 32 (1998), pp. 1972–1979.

Westall, J.C.: Reactions at the oxide-solution interface: chemical and electrostatic models. Geochemical Processes at Mineral Surfaces. In: J. Davis & K.F. Hayes: *Geochemical Processes at Mineral Surfaces (ACS Symp. Ser. 323)*, 1986, pp. 54–78.

Wilkin, R.T., Puls, R.W. & Sewell, G.W.: *Long-term performance of permeable reactive barriers using zero-valent iron: an evaluation at two sites.* USEPA Environmental Research Brief, Cincinnati, 2002, http://nepis.epa.gov/Exe/ZyNET.exe/30003TJ7.TXT?ZyActionD=ZyDocument&Client=EPA&Index=2000+Thru+2005&Docs=&Query=&Time=&EndTime=&SearchMethod=1&TocRestrict=nn&Toc=&TocEntry=&QField=&QFieldYear=&QFieldMonth=&QFieldDay=&IntQFieldOp=0&ExtQFieldOp=0&XmlQuery=&File=D%3A%5Czyfiles%5CIndex%20Data%5C00thru05%5CTxt%5C00000006%5C30003TJ7.txt&User=ANONYMOUS&Password=anonymous&SortMethod=h%7C-&MaximumDocuments=1&FuzzyDegree=0&ImageQuality=r75g8/r75g8/x150y150g16/i425&Display=p%7Cf&DefSeekPage=x&SearchBack=ZyActionL&Back=ZyActionS&BackDesc=Results%20page&MaximumPages=1&ZyEntry=1&SeekPage=x&ZyPURL (accessed May 2012).

Wilkin, R.T., Puls, R.W. & Sewell, G.W.: Long-term performance of permeable reactive barriers using zero-valent iron: geochemical and microbiological effects. *Ground Water* 41 (2003), pp. 493–503.

Wilkin, R.T., Acree, S.D., Ross, R.R., Lee, T.R. & Beak, D.G.: An *in-situ* permeable reactive barrier for the treatment of arsenic in ground water. *Presented at Geological Society of America Annual Conference*, 22–25 October, 2006, Philadelphia, PA, 2006.

Wilkin, R.T., Acree, S.D., Ross, R.R., Beak, D.G. & Lee T.R.: Performance of a zerovalent iron reactive barrier for the treatment of arsenic in groundwater: Part 1. Hydrogeochemical studies. *J. Contam. Hydrol.* 106 (2009), pp. 1–14.

Xiong, Z., He, F., Zhao, D. & Barnett, M.O.: Immobilization of mercury in sediment using stabilized iron sulfide nanoparticles. *Water Res.* 43 (2009), pp. 5171–5179.

Xu Y. & Zhang, W.X.: Subcolloidal Fe/Ag particles for reductive dehalogenation of chlorinated benzenes. *Ind. Eng. Chem. Res.* 39 (2000), pp. 2238–2244.

Yang, L., Donahoe, R.J. & Redwine, J.C.: In situ chemical fixation of arsenic-contaminated soils: an experimental study. *Sci. Total Environ.* 387 (2007), pp. 28–41.

Younger, P.L., Banwart, S.A. & Hedin, R.S.: *Minewater: hydrology, pollution and remediation.* Kluwer Academic Publishers, Dordrecht, The Netherlands, 2002.

Yuan, C. & Chiang, T.-S.: The mechanisms of arsenic removal from soil by electrokinetic process coupled with iron permeable reaction barrier. *Chemosphere* 67 (2007), pp. 1533–1542.

Yuan, C. & Chiang, T.-S.: Enhancement of electrokinetic remediation of arsenic spiked soil by chemical reagents. *J. Hazard. Mater.* 152 (2008), pp. 309–315.

Zelmanov, G. & Semiat, R.: Iron(3) oxide-based nanoparticles as catalysts in advanced organic aqueous oxidation. *Water Res.* 42 (2008), pp. 492–498.

Zhan, J., Zheng, T., Piringer, G., Day, C., McPherson, G.L., Lu, Y., Papadopoulos, K. & John, V.T.: Transport characteristics of nanoscale functional zerovalent iron/silica composites for in situ remediation of trichloroethylene. *Environ. Sci. Technol.* 42 (2008), pp. 8871–8876.

Zhan, J., Sunkara, B., Le, L., John, V.T., He, J., McPherson, G.L., Piringer, G. & Lu, Y.: Multifunctional colloidal particles for in situ remediation of chlorinated hydrocarbons. *Environ. Sci. Technol.* 43 (2009), pp. 8616–8621.

Zhan, J., Kolesnichenko, I., Sunkara, B., He, J., McPherson, G.L., Piringer, G.& John, V.T.: Multifunctional iron-carbon nanocomposites through an aerosol-based process for the in situ remediation of chlorinated hydrocarbons. *Environ. Sci. Technol.* 45 (2011), pp. 1949–1954.

Zhang, M., He, F., Zhao, D. & Hao, X.: Degradation of soil-sorbed trichloroethylene by stabilized zero valent iron nanoparticles: effects of sorption, surfactants, and natural organic matter. *Water Res.* 45 (2011), pp. 2401–2414.

Zhang, W. Nanoscale iron particles for environmental remediation: An overview. *J. Nanoparticle Res.* 5, 2003, pp. 323–332.

Zhang W.-X., Wang, C.-B. & Lien, H.-L.: Treatment of chlorinated organic contaminants with nanoscale bimetallic particles. *Catal. Today* 40 (1998), pp. 387–395.

Zhang, Z., Li, M., Chen, W., Zhu, S., Liu, N. & Zhu, L.: Immobilization of lead and cadmium from aqueous solution and contaminated sediment using nano-hydroxyapatite. *Environ. Pollut.* 158 (2010), pp. 514–519.

Zhao, X., Guo, X., Yang, Z., Liu, H. & Qian, Q.: Phase-controlled preparation of iron (oxyhydr)oxide nanocrystallines for heavy metal removal. *J. Nanopart. Res.* 13 (2011), pp. 2853–2864.

Zheng, T., Zhan, J., He, J., Day, C., Lu, Y., McPherson, G.L., Piringer, G. & John, V.T.: Reactivity characteristics of nanoscale zerovalent iron-silica composites for trichloroethylene remediation. *Environ. Sci. Technol.* 42 (2008), pp. 4494–4499.

Zouboulis, A.I. & Katsoyiannis, I.A.: Recent advances in the bioremediation of arsenic-contaminated groundwaters. *Environ. Intern.* 31 (2005), pp. 213–219.

CHAPTER 2

Numerical modeling of arsenic mobility

Ilka Wallis, Henning Prommer & Dimitri Vlassopoulos

2.1 INTRODUCTION

Assessment and successful remediation of arsenic contaminated sites requires a rigorous understanding of the factors influencing arsenic fate and transport as well as the ability to predict the behavior of arsenic in soil and aquifer systems under future conditions. The fate of arsenic in both naturally and anthropogenically impacted aquifer systems, however, is determined by interactions between physical, chemical, and biological processes. In many cases it can be difficult to discern or distinguish between specific geochemical or biogeochemical processes controlling the fate of arsenic, or whether its fate in fact might be predominantly controlled by physical transport. The underlying causes of elevated arsenic concentrations will generally vary from case to case and can be affected by several site-specific factors including the nature and magnitude of historical or ongoing releases, the chemical and mineralogical composition of the aquifer matrix, the composition of the ambient groundwater and hydrogeological conditions.

Due to the complexity of the interactions between hydrologic, chemical, and biogeochemical processes which are generally nonlinear, arsenic fate can often be non-intuitive and it becomes obvious that analyses and predictions based on single-solute transport models are overly simplistic and of limited value. Therefore numerical models that integrate these processes, specifically geochemical and reactive transport models, provide an important pathway to asses and quantify biogeochemical processes. Over the last two decades continuous advances in computational performance have facilitated an increasing use of numerical simulators that fully couple both geochemical and hydrological processes. Based on knowledge derived from either theoretical work, laboratory investigations and/or site-specific field work, these numerical models provide a quantitative framework to evaluate arsenic fate in complex subsurface systems, thereby allowing simulation of time- and space-varying flow, transport and biogeochemical reactions.

In this chapter we provide a brief overview of the approaches and mathematical descriptions that are used to simulate and quantify the processes affecting As behavior in aquifers.

2.2 MODELLING APPROACHES, TYPES OF MODELS AND COMMON MODELLING TOOLS

Analysis and prediction of arsenic behavior in groundwater may be carried out over different time and spatial scales at widely differing levels of geochemical complexity. Therefore, the appropriate choice of modeling tools and approaches can vary considerably, depending on the nature and objectives of the study as well as the types and amount of data that are available to constrain the models. Prior to the use of numerical models (i) the clear definition of the modeling objectives and (ii) the development of preliminary conceptual models of the chemistry and chemical changes within the groundwater systems are important prerequisite that cannot be overemphasized. In the early phases of model development, geochemical indicators provide information on groundwater provenance, flow directions and rates, and age which form the basis of conceptual hydrogeological models. *Vice versa*, knowledge of aquifer hydrostratigraphy, predominant groundwater flow patterns and similar types of information help to develop conceptual models of hydrogeochemical processes.

Unraveling observed water quality patterns in aquifers can be a daunting task. The input of different sources and types of water is the first of factors that adds to this complexity. Sources include precipitation, surface waters, seawater, ascending deep groundwater and anthropogenic sources such as wastewater. The overprint of geochemical processes adds to the complexity since the water composition is altered as it travels through the subsurface. Mineral dissolution and precipitation, ion exchange and redox reactions are generally the most important chemical processes that affect groundwater quality. Mixing of different water types through physical processes such as advection, dispersion and diffusion also exerts an influence on the composition of groundwater.

Traditional methods that are applied to interpret large hydrochemical data sets involve plotting (Fetter, 2001) and classification of samples into groups (e.g., Stuyfzand, 1993) in order to be able to discern regional trends and to identify chemical processes. As useful as these visual and statistical methods can be in deriving preliminary conceptual models for the reactions that give rise to the range of observed groundwater compositions, they are incapable of determining whether the inferred reactions are thermodynamically feasible. Geochemical and reactive transport models that quantitatively simulate reactions subject to the laws of thermodynamics are therefore a natural choice for confirming whether a particular conceptual model is also chemically realistic. Or, as Lichtner (1996) suggested: 'Quantitative models force the investigator to validate or invalidate ideas by putting real numbers into an often vague hypothesis and thereby starting the thought process along a path that may result in acceptance, rejection, or modification of the original hypothesis'.

The first quantitative geochemical models were originally developed in the 1960s, initially with the sole objective of calculating the speciation of dissolved ions. These original models were soon improved and extended to include chemical reactions that alter the water composition such as mineral equilibria and cation exchange (e.g., Parkhurst, 1980; Plummer, 1983). Geochemical modeling plays an important role as a stand-alone tool for understanding controls on water chemistry, but is also a key part of the conceptual model development and testing for coupled reactive transport models. In the context of groundwater arsenic problems, typical questions and tasks that can be addressed with geochemical models can include:

- What are the predominant arsenic species (complexes) present in groundwater?
- Is groundwater under- or over-saturated or at equilibrium with respect to specific minerals?
- Which reactant(s) could be added to groundwater of a specific composition to reduce dissolved arsenic concentrations?
- How does arsenic partition between aqueous and sorbed phases?
- Could changes in water composition and/or pH lead to arsenic release from or uptake by the aquifer solids?

There are several commonly used public domain and/or commercially available software for carrying out these tasks including PHREEQC, MINTEQA2 and the Geochemist's Workbench (see Table 2.1 for more details).

The construction of numerical groundwater flow models on the other hand can provide an in-depth analysis of the flow system and is a prerequisite for subsequent solute and reactive transport modeling endeavors. Numerical models are necessary especially for simulation of systems characterized by geological heterogeneities and/or complex transient hydrologic forcings. Typical questions that can be explored with numerical groundwater flow models include:

- What are the flow velocities in the vicinity and particularly downstream of an arsenic pollution source?
- Could an arsenic plume be contained by a specific pumping well configuration?

The USGS groundwater model MODFLOW (Harbaugh, 2005; Harbaugh *et al.*, 2000) is the most widely used code for simulation of multi-dimensional flow fields, mostly owing to the fact that its open-source code is well documented. Other frequently used codes include HST3D (Kipp, 1986), FEMWATER (Yeh *et al.*, 1992) and FEFLOW (Diersch, 1997) (Table 2.1).

Based on these (underlying) groundwater flow simulators, advective-dispersive transport codes such as MT3DMS (Zheng and Wang, 1999), FEFLOW (Diersch, 1997), or FEMWATER (Yeh *et al.*, 1992) allow for the simulation of multi-dimensional solute transport behavior due to physical transport processes (i.e., advection, dispersion, diffusion), often also in conjunction with simplified representations of reactive processes such as linear and non-linear sorption and decay. The application of such codes to arsenic problems can sometimes be justified, for example in aquifers where geochemical conditions remain more or less stable in space and time and thus the variations in arsenic transport characteristics are less pronounced. However, these solute transport codes do not account for coupling and interactions between various dissolved species and reactive mineral phases, which often strongly impact arsenic mobility.

Simulation of these more complex reactive transport systems can be achieved through multi-component reactive transport simulators. Such codes often couple a geochemical model (such as PHREEQC) with a solute transport simulator to model changes in water composition due to both physical transport and geochemical reactions as it travels through the subsurface. Significant advances in computational power, numerical solution techniques and parallelization of codes have allowed routine simulation of increasingly complex groundwater quality problems in heterogeneous aquifer systems where flow, transport and reactive processes need and can be considered simultaneously.

Table 2.1.   Widely used geochemical, flow, solute transport and reactive transport modeling tools.

Code	Website	Summary
A: Geochemical models		
PHREEQC	http://wwwbrr.cr.usgs.gov/projects/GWC_coupled/phreeqc/	PHREEQC (version 3) is designed to perform a wide variety of aqueous geochemical calculations. PHREEQC implements several types of aqueous models: two ion-association aqueous models (the Lawrence Livermore National Laboratory model and WATEQ4F), a Pitzer specific-ion-interaction aqueous model, and the SIT (Specific ion Interaction Theory) aqueous model. Using any of these aqueous models, PHREEQC has capabilities for (1) speciation and saturation-index calculations; (2) batch-reaction and one-dimensional (1D) transport calculations with reversible and irreversible reactions, which include aqueous, mineral, gas, solid-solution, surface-complexation, and ion-exchange equilibria, and specified mole transfers of reactants, kinetically controlled reactions, mixing of solutions, and pressure and temperature changes; and (3) inverse modeling.
Geochemists Workbench	http://www.gwb.com/	Geochemist's Workbench is an integrated set of interactive software tools for solving a range of problems in aqueous geochemistry including balancing chemical reactions, speciation calculations, calculating stability diagrams and the equilibrium states of natural waters, tracing reaction processes, modeling reactive transport in one and two dimensions, and plotting the results of these calculations. Heat transfer by advection and conduction, mineral precipitation and dissolution, complexation and dissociation, sorption and desorption, redox transformations and reactions catalyzed by microbial activity can be simulated.

(*continued*)

Table 2.1.   Continued.

Code	Website	Summary
MINTEQA2	http://www2.epa.gov/exposure-assessment-models/minteqa2	MINTEQA2 is a geochemical equilibrium speciation model capable of computing equilibria among dissolved, adsorbed, solid, and gas phases. MINTEQA2 includes an extensive database of thermodynamic data for many constituents of environmental interest.

B: Single- and multi-species transport models

Code	Website	Summary
FEFLOW	http://www.feflow.com/aboutfeflow.html	FEFLOW is a finite-element software package for modeling fluid flow and transport of dissolved constituents and/or heat transport processes in the subsurface. Multi-dimensional groundwater flow, including density-dependent flow can be combined with the simulation of multi-species solute transport in groundwater and the unsaturated zone. Reaction modules available include decay and sorption and the possibility to define kinetic reactions via an equation editor.
FEMWATER	http://homepage.usask.ca/~mjr347/gwres/femwref.pdf	A 3D finite-element model used to simulate density-driven coupled flow and contaminant transport in saturated and unsaturated zones. FEMWATER allows modeling of salinity intrusion and other density-dependent contaminants and simple transport problems such as sorption and decay.
HST3D	http://wwwbrr.cr.usgs.gov/projects/GW_Solute/hst/	The HST3D code is suitable for simulating groundwater flow and associated heat and solute transport in saturated, three-dimensional flow systems with variable density and viscosity.
HYDRUS-1D/2D/3D	http://igwmc.mines.edu/software/hydrus.html	HYDRUS is a software package for the analysis of water flow and solute transport in variably saturated porous media. The geochemical UNSATCHEM module (Šimůnek and Suarez, 1994; Šimůnek et al., 1996) has been implemented into the HYDRUS (2D/3D) software package to simulate the transport of major ions in variably-saturated porous media, including major ion equilibrium and kinetic non-equilibrium chemistry. The model accounts for various equilibrium chemical reactions between these components, such as complexation, cation exchange and precipitation-dissolution.
MT3DMS	http://hydro.geo.ua.edu/mt3d/	The multi-species transport simulator MT3DMS has the capability of modeling changes in concentrations of groundwater contaminants due to advection, dispersion, diffusion, and some chemical reactions including equilibrium-controlled linear or non-linear sorption, and first-order irreversible or reversible kinetic reactions. MT3D/MT3DMS uses the output head and cell-by-cell flow data computed by MODFLOW to establish the groundwater flow field.

(*continued*)

Table 2.1. Continued.

Code	Website	Summary
RT3D	http://bioprocess. pnnl.gov/	RT3D (Clement, 1997) is a reactive transport code based on MT3DMS. It extends the capabilities of MT3DMS to solve arbitrary kinetic reaction problems. RT3D provides a number of predefined reaction packages, e.g., for biodegradation of oxidizable contaminants consuming one or more electron acceptors and for sequential decay chain-type reactions of chlorinated hydrocarbons (CHCs).
SUTRA	http://water.usgs.gov/ nrp/gwsoftware/sutra/ sutra.html	SUTRA (Saturated-Unsaturated Transport) is a computer program that simulates fluid density-dependent saturated or unsaturated groundwater flow and solute transport. Solute transport includes conservative transport, equilibrium adsorption, first-order and zero-order production or decay and transport of thermal energy in the groundwater and solid matrix of the aquifer.

C: Coupled reactive transport models

Code	Website	Summary
CrunchFlow	www.csteefel.com	CrunchFlow is a code for multicomponent reactive flow and transport, which can be used for simulation of a range of processes, including reactive contaminant transport, chemical weathering, carbon sequestration, biogeochemical cycling, and water-rock interaction. It features a wide range of biogeochemical processes and has integrated flow and solute transport components.
HBGC123D	http://hbgc.emsgi. org/main.html	HBGC123D is a computer code for the simulation of coupled hydrologic transport and biogeochemical kinetic and/or equilibrium reactions in variably saturated media.
HP2	http://www.pc-progress.com/en/ Default.aspx?h3d2-HP2	HP2 (acronym for HYDRUS-PHREEQC-2D), is a computer code, which couples Hydrus-2D/3D with the PHREEQC geochemical code (Parkhurst and Appelo, 1999). HP2 contains modules simulating (1) transient water flow, (2) the transport of multiple components, (3) mixed equilibrium/kinetic biogeochemical reactions, and (4) heat transport in two-dimensional variably-saturated porous media (soils). The coupled program can simulate a broad range of low-temperature biogeochemical reactions in water, the vadose zone and in groundwater systems, including interactions with minerals, gases, exchangers and sorption surfaces based on thermodynamic equilibrium, kinetic, or mixed equilibrium-kinetic reactions.
MIN3P	http://www.eos.ubc. ca/research/hydro/ research/min3p/ reactive_tran_web.htm	MIN3P is a flow and reactive transport code for variably saturated media. Advective-diffusive transport in the water phase and diffusive transport in the gas phase are included. The reaction network is designed to handle kinetic and equilibrium controlled reactions, dissolution-precipitation

(continued)

Table 2.1.   Continued.

Code	Website	Summary
		reactions, aqueous complexation, gas partitioning between phases, oxidation-reduction, ion exchange, and surface complexation.
PFLOTRAN	http://www.pflotran. org/	PFLOTRAN is an open source subsurface flow and reactive transport code. PFLOTRAN is used to describe multiphase, multicomponent and multiscale reactive flow and transport in porous materials. The code is specifically designed to run on massive parallel computing architectures and includes adaptive mesh refinement.
PHAST	http://wwwbrr.cr.usgs. gov/projects/GWC_ coupled/phast/	PHAST is a 3-dimensional, reactive-transport simulator that is available for Windows and Linux. PHAST simulates constant-density saturated flow, multicomponent transport, and a wide range of equilibrium and kinetic chemical reactions.
PHT3D	http://www.pht3d.org	PHT3D is a three-dimensional reactive multicomponent transport code for saturated porous media. PHT3D couples the transport simulator MT3DMS and the geochemical model PHREEQC-2. It handles a wide range of mixed equilibrium/kinetic geochemical reactions.
STOMP	http://www.stomp. pnl.gov	Developed at the Pacific Northwest National Laboratory (PNNL) as a suit of numerical simulators for multifluid subsurface flow and transport processes. It includes the module ECKEChem, which is the reactive transport package for the STOMP simulator and an algorithm for equilibrium, conservation, and kinetic equation chemistry. The principal application objective for the ECKEChem capabilities is the scalable simulation of the geochemical reactions associated with injection CO_2 into geologic reservoirs.
TOUGH REACT	http://esd.lbl.gov/ research/projects/tough/	TOUGH REACT is a numerical simulation program for chemically reactive non-isothermal flows of multiphase fluids in porous and fractured media, developed by introducing reactive chemistry into the multiphase flow code TOUGH2. Interactions between mineral assemblages and fluids can occur under local equilibrium or kinetic rates. The gas phase can be chemically active. Precipitation and dissolution reactions can change formation porosity and permeability, and can also modify the unsaturated flow properties of the rock.

2.3   THE SIMULATION OF PROCESSES AFFECTING AS TRANSPORT BEHAVIOR

2.3.1   *Modeling groundwater flow and solute transport*

The successful application of reactive transport models to simulate the fate of arsenic in aquifers relies strongly on the accurate representation of the underlying groundwater flow characteristics such as flow velocities and their variations in space and time and, linked to that, hydrodynamic mixing and dilution effects. In cases where active remediation requires addition of specific amendments to the aquifer, the rate of delivery of such reactants via advective-dispersive transport can often be the limiting factor in the progress of *in-situ* treatment. In such cases developing a good understanding (and associated model) of the groundwater flow processes, subsurface mixing of multiple solutes induced by hydrodynamic dispersion, and contact times between solutes and minerals, can be key to successful remediation. A detailed understanding and quantification of the natural groundwater dynamics as well as the dynamics induced by the extraction and/or injection itself are essential for optimizing engineered schemes or predicting natural attenuation rates.

An equation describing the transport and dispersion of a single, dissolved conservative species in flowing groundwater within a homogeneous porous medium may be derived based on the principle of mass balance (e.g., Domenico and Schwartz, 1998). The mass balance statement requires that the change in mass storage of a species within a representative elementary aquifer volume (REV) during a given time interval is equal to the difference in the mass inflows and outflows due to dispersion, advection and external sinks and sources over the same interval. Mathematically, this mass balance is most frequently described by the advection-dispersion equation (ADE) (Bear, 1972; Bear and Verruijt, 1987; Zheng and Wang, 1999):

$$\frac{\partial C}{\partial t} = \frac{\partial}{\partial x_i}\left(\mathbf{D}_{ij}\frac{\partial C}{\partial x_j}\right) - \frac{\partial}{\partial x_i}(v_i C) + \frac{q_s}{\theta}C_q \qquad (2.1)$$

where C [ML^{-3}] is the dissolved concentration of a chemical species, v_i [LT^{-1}] is the seepage or linear pore water velocity in direction x_i [L], $\mathbf{D}_{ij}$ [L^{2}T^{-1}] is the hydrodynamic dispersion coefficient tensor (summation convention assumed), θ is the porosity of the subsurface medium, q_s [L^{3}L^{-3}T^{-1}] is the volumetric flow rate per unit volume of water representing external fluid sources and sinks and C_q [ML^{-3}] is the concentration of a species within this flux if q_s is positive (injection), otherwise $C_q = C$.

The first term on the right side of the ADE represents the rate of change in concentration due to dispersion, whereby the dispersion term represents two processes: molecular diffusion as well as mechanical dispersion. Mechanical dispersion results from the microscopic fluctuation of streamlines in space with respect to the mean groundwater flow direction and small-scale changes of porosity and hydraulic conductivity. Mechanical dispersion is assumed to be scale-independent within the ADE and often initially estimated, for example, on the basis of literature values and subsequently optimized during the model calibration for which concentration data of conservative solutes most commonly serve as calibration constraints (e.g., Fiori and Dagan, 1999; Greskowiak *et al.*, 2005; Jensen *et al.*, 1993; Wallis *et al.*, 2010). This pragmatic approach appears in many cases to produce a reasonable description of solute spreading and mixing behavior, especially for local-scale problems and for aquifers that are relatively homogeneous. However, it should be noted that this approach can be too simplistic and, while perhaps still capturing the general patterns, may sometimes not satisfactorily reproduce the spreading behavior observed at larger scales (Konikow, 2010). This will also depend on the level of detail at which larger-scale geological features and structures and their impact on groundwater flow are represented within the numerical model.

The second term on the right side of the ADE describes advective transport. Thereby it is assumed that the dissolved species is transported at the same mean velocity as the flowing groundwater, which, in field-scale contaminant problems within the saturated groundwater zone, does represent the dominant physical transport process. The average seepage velocity of the flowing groundwater, i.e., v_i in Equation (2.1), is derived from Darcy's law and the three-dimensional flow equation for saturated groundwater.

The last term on the right hand side of the ADE represents the effects of mixing of waters from external sources and sinks with different solute concentrations. For non-conservative solute transport, additional terms can be added to the right side of the ADE to account for geochemical reactions, such as sorption, precipitation/dissolution of minerals, decay etc. The geochemical reactions most relevant to arsenic transport in groundwater and approaches for representing these within a numerical modeling framework are described in the following sections.

The coupled hydrologic transport and geochemical reaction equations can only be solved analytically for some very simple cases. For more complicated cases, e.g., involving heterogeneous aquifers, transient boundary conditions, etc., most solution procedures are based upon numerical techniques such as the finite difference and finite element methods and the reader is referred to modeling-specific texts such as Anderson and Woessner (1992), Chiang and Kinzelbach (2000) and Zheng and Bennett (2005) for a detailed description of these techniques.

2.3.2 *Processes controlling the geochemical environment*

To a large extent, the geochemical changes in a hydrogeological system are controlled by microbially mediated redox reactions (e.g., Champ 1979; Christensen *et al.*, 2000; Eckert and Appelo, 2002; Massmann *et al.*, 2004). In many groundwater systems, these redox reactions are often driven by the mineralization of naturally occurring sediment-bound and/or dissolved organic matter (e.g., Hunter *et al.*, 2008; Kirk *et al.*, 2004; Park *et al.*, 2009). However, in contaminated aquifers, these biogeochemical redox reactions may also be driven by mineralization of xenobiotic substances such as petroleum hydrocarbons or pesticides that may co-occur with arsenic (e.g., Baedecker *et al.*, 1993; Levine *et al.*, 1997). Mineralization of organic substances involves the consumption of electron acceptors such as oxygen, nitrate, manganese- and iron-(hydro)oxides and sulfate (e.g., Chapelle and Lovley, 1992; McMahon and Chapelle, 1991). The consumption of the electron-acceptors typically occurs in a sequential order constrained by thermodynamic principles (e.g., Christensen *et al.*, 2000; Postma and Jakobsen, 1996; Stumm and Morgan, 1996). This usually results in the formation of spatially distinct redox zones along the groundwater flow direction which range typically from aerobic, to denitrifying, Mn- and Fe-reducing, sulfate reducing and methane producing conditions.

Due to its variable oxidation states the mobility of arsenic is strongly influenced by the redox environment, which determines sorption characteristics and therefore the transport behavior of As (e.g., Goldberg, 2002). For example, arsenate (As(V)), the predominant chemical species under aerobic conditions, tends to be more strongly adsorbed onto oxide mineral surfaces such as ferrihydrite and thus is typically largely immobile under near-neutral pH conditions. Conversely, arsenite (As(III)), the prevalent chemical species under reducing conditions, adsorbs less strongly and is generally more mobile.

Biogeochemical redox reactions often induce additional secondary reactions such as precipitation and dissolution of minerals, ion exchange or surface complexation reactions, which can also greatly influence the mobility of arsenic (e.g., Appelo and de Vet, 2002). Therefore, efforts to simulate As behavior in the subsurface should incorporate the processes accounting for the spatial and temporal variations of the redox chemistry of the aquifer.

Within reactive transport models, different approaches can be applied to simulate organic substrate oxidation. The most commonly applied reaction rate formulations are based on the application of Monod kinetics (Barry *et al.*, 2002). A relatively simple variant of Monod-type rate expressions for the kinetically controlled oxidation of organic matter by oxygen, nitrate or sulfate is (Parkhurst and Appelo, 1999; Prommer *et al.*, 2006, Sharma *et al.*, 2012):

$$r_{\text{DOC}} = \left(k_{O_2} \frac{C_{O_2}}{2.94 \times 10^{-4} + C_{O_2}} + k_{NO_3^-} \frac{C_{NO_3^-}}{1.55 \times 10^{-4} + C_{NO_3^-}} + k_{SO_4^{2-}} \frac{C_{SO_4^{2-}}}{1.0 \times 10^{-4} + C_{SO_4^{2-}}} \right)$$

(2.2)

where r_{DOC} is the overall degradation rate of organic matter, k_{O_2}, $k_{NO_3^-}$, $k_{SO_4^{2-}}$ are the rate constants of carbon mineralization under aerobic, denitrifying and sulfate-reducing conditions and C_{O_2}, $C_{NO_3^-}$, $C_{SO_4^{2-}}$ are the concentrations of oxygen, nitrate and sulfate in the groundwater.

The geochemical response of the organic substrate mineralization, i.e., the ensuing redox reactions, is typically simulated either via a single-step or by a two-step approach (Brun and

Engesgaard, 2002). In the single-step approach, a redox reaction is modeled as an irreversible kinetic reaction that simultaneously includes organic substrate oxidation and the reduction of a predefined electron acceptor (such as oxygen or nitrate). In contrast, the two-step approach or so-called partial equilibrium approach (PEA) separates the electron donating step (e.g., organic matter degradation) from the electron accepting step. The first step is assumed to be the rate-limiting step (kinetically controlled), and thus the second step can be simply modeled as an equilibrium (instantaneous) reaction (Postma and Jakobsen, 1996; Prommer *et al.*, 2002). Using the latter approach, redox reactions driven by organic matter degradation can be simulated using e.g., PHREEQC-2 by simply adding inorganic carbon at the same rate (i.e., r_{DOC}, Equation (2.2)) at which the organic substrate is removed. Internally, redox equilibrium may be assumed among all electron acceptors, which will then be automatically consumed in the order of their thermodynamic favorability without the need to explicitly define specific electron accepting reactions and their order of consumption. In contrast, alternative modeling approaches may assume and model the redox reactions as a single, coupled reaction (e.g., Hunter *et al.*, 1998; MacQuarrie and Sudicky, 2001). Besides organic matter, pyrite and/or other reduced minerals may also affect the redox zonation within the aquifer. Examples are for instance areas of elevated nitrate due to recharge from agricultural sources due to overuse of fertilizers (Zhang *et al.*, 2009). Here, pyrite oxidation may lead to acidification of the groundwater, if the aquifer contains insufficient amounts of minerals that can act as pH buffers, such as calcite. Moreover, pyrite oxidation has often been shown to be accompanied by the release of trace metals, including arsenic (e.g., Price and Pichler, 2006; Smedley and Kinniburgh, 2001; Zhang *et al.*, 2009).

2.3.3 *Sorption and desorption*

Once arsenic is mobilized, adsorption-desorption reactions are important processes that subsequently exert a major control on the effective transport rates of arsenic and on whether its mobility becomes problematic for downstream receptors. A range of mathematical expressions for arsenic adsorption-desorption reactions have been used in reactive transport models, from relatively simple empirical relationships, such as distribution coefficients and isotherm equations (Freundlich, 1926; Langmuir, 1918) to the more sophisticated surface complexation modeling approaches (Appelo and Postma, 2005; Goldberg *et al.*, 2007).

When adsorption is represented by empirical approaches, the conceptual model for adsorption is generally simplistic in that it ignores potential effects of variable chemical conditions, such as changes in pH or solute concentration spatially or with time on the partitioning behavior (Davis and Kent, 1990). The simplest empirical relationship for adsorption used in reactive transport models is the linear equilibrium sorption model:

$$S = K_d \times C \qquad (2.3)$$

where K_d is the distribution coefficient [L^3/M], C is the concentration of the dissolved phase [M/L^3] and S is the mass of the solute species (arsenic) adsorbed on the aquifer solids [M/M].

The so-called K_d approach then describes a temporally constant ratio between dissolved solute concentration in the groundwater and the concentration sorbed to the sediment, i.e., the latter is linearly proportional to the former. The proportionality constant (K_d) or slope of the isotherm relates to the retardation that the solute experiences relative to a non-reactive tracer during groundwater flow (Appelo and Postma, 2005). Note that in this formulation the adsorption capacity of the aquifer is assumed to be infinite. This assumption is, however, not necessarily suitable for simulating arsenic, since the number of sorption sites that can be occupied by arsenic and other solutes that compete for these sites is often limited.

Somewhat more complex empirical relationships, i.e., Freundlich and/or Langmuir adsorption isotherms are often employed to account for the finite adsorption capacity of the aquifer media in describing the relationship between sorbed and dissolved solute concentrations. The Freundlich isotherm has the form:

$$S = K_F \times C^n \qquad (2.4)$$

where K_F [L^3/M] and n [−] are empirical coefficients and S and C are defined as above.

If the coefficient n equals 1, the Freundlich isotherm reduces to the linear sorption model. However, n is usually selected to be smaller than 1, so that the increase of sorbed concentrations decreases as the solute concentration increases. As is the case of the linear sorption model, no upper adsorption limit is simulated and sorption extends infinitely as dissolved concentrations increase.

In contrast the Langmuir adsorption equation introduces an upper limit to adsorption:

$$S = \frac{QKC}{1 + KC} \qquad (2.5)$$

where K is the partition coefficient reflecting the extent of sorption, Q is the maximum sorption capacity [M/M] and C is defined as above.

In this description, sorption still increases linearly at low dissolved solute concentrations. However, as aqueous concentrations increase further, the distribution coefficient, i.e., the ratio between the sorbed and dissolved concentration decreases gradually as surface sites become increasingly occupied. Eventually the isotherm flattens out as the sorption sites become fully saturated.

These empirical formulations assume that the ambient solution concentration is at equilibrium with the sorbed phase concentration. However, the equilibrium assumption may not be valid during transport, in which case kinetic expressions may be applied to describe the time-dependence of adsorption-desorption process. The simplest kinetic expression of sorption is a first-order reversible reaction:

$$\rho_b \frac{\partial S}{\partial t} = \beta \left(C - \frac{S}{K_d} \right) \qquad (2.6)$$

where ρ_b is the bulk density of the subsurface medium [M/L^3], β is the first-order mass transfer rate [1/T] between the liquid and solid phases. Kinetic expressions describing nth-order reversible adsorption have also been described (Darland and Inskeep, 1997).

The above mentioned empirical sorption models require knowledge of the mineral composition and are dependent on the observed relationships between aqueous and sorbed concentrations. Sorption isotherms are generally measured in the laboratory using batch equilibrium or column breakthrough experiments and these results are then applied within a solute transport model to predict field-scale transport behavior. However, these empirically derived partitioning relationships are only valid for the conditions under which they were determined and do not account for potential impacts of spatial or temporal variability in mineralogy and/or hydrochemistry on adsorption. Therefore empirical adsorption models are generally only appropriate when applied under well-controlled laboratory conditions or where field-sites are characterized by relatively constant geochemical conditions.

There are numerous examples where standard adsorption isotherm equations have been fitted to replicate observed laboratory data sets on arsenic sorption onto a variety of sorbents (e.g., Hsia *et al.*, 1992; Masue *et al.*, 2007; Nath, 2009; Wolthers, 2005). Such calculation can be performed by batch-type geochemical models (see Table 2.1).

Applications of numerical modeling studies in which empirical relationships have been used to represent arsenic sorption behavior are also manifold. For example, work undertaken by DPHE *et al.* (1999) combined results from groundwater flow models with estimates derived from sorption isotherms to assess and predict the rate of arsenic movement in typical Bangladesh aquifers. Darland and Inskeep (1997) studied the applicability of equilibrium adsorption models (linear and Freundlich) and kinetic adsorption models (first-order and nth-order reversible) for simulating arsenic transport under varying pore water velocities. Decker *et al.* (2006) developed a modified formulation of the Langmuir isotherm to simulate pH-dependent sorption behavior of both As(III) and As(V). The formulation was incorporated into the variable saturated reactive transport solver UNSATCHEM (Simunek and Suarez, 1994; Simunek *et al.*, 1996), to predict the potential for As impacts on groundwater from gold mining-related activities. More recently Jeppu and Clement (2012) introduced a modified sorption isotherm to simulate pH-dependent adsorption. The modified Langmuir-Freundlich (MLF) isotherm was used to successfully describe the sorption behavior of arsenic on two sorbents, i.e., goethite and goethite-coated sand.

While in the above-mentioned examples the empirical formulations were able to successfully describe the observed arsenic sorption behavior, it is important to note that their application is strictly only warranted under the geochemical conditions for which they were originally developed. In natural systems, the number of factors influencing the sorption efficiency of arsenic may be large. Besides pH, arsenic concentration and the aquifer redox status, the concentration of competing solutes such as phosphate, silicate or bicarbonate or feedback mechanisms between ambient chemistry and the changes induced by the adsorption reaction itself may affect sorption and it remains almost impossible to capture those impacts by relatively simple model approaches (Goldberg, 2007). Thus, for field sites which exhibit spatial or temporal variability in geochemical conditions, surface complexation models (SCMs) are typically used to describe the interactions of arsenic with sorbing surfaces and the factors which impact these interactions. The SCMs allow arsenic adsorption behavior to change as a function of varying geochemical conditions in space and time. Of course the corresponding increase in model complexity comes at the price of an increased computational demand and, perhaps more significantly, additional input data requirements.

The most influential study in this context was probably the work of Dzombak and Morel (1990). They reviewed the available laboratory data for the adsorption of arsenic (and other ions) by Fe-oxides and fitted the most reliable data with a diffuse double-layer SCM. This model, while developed for adsorption to pure minerals and therefore ignoring the heterogeneity that is present in natural groundwater systems, has been applied within many purely geochemical, i.e., batch-type, as well as reactive transport modeling studies. One of the prominent reasons for its widespread use is that the SCM and the accompanying thermodynamic database are readily available within most of the earlier discussed geochemical models. However, other complexation models have also been proposed, including the constant capacitance model (Stumm *et al.*, 1980), the triple layer model (Davis *et al.*, 1978) and the CD-MUSIC model (Hiemstra and van Riemsdijk, 1999). Detailed appraisals of the various SCM models can be found in the review of Goldberg *et al.* (2007) and Appelo and Postma (2005).

Surface complexation models are generally regarded as a progression from empirical formulations, as they incorporate a more fundamental process-based representation of sorption and desorption and allow for direct coupling of sorption and geochemical equilibria via a thermodynamic database (Payne *et al.*, 2013). However, the most appropriate methods to simulate surface complexation, especially in natural systems, and the methods to determine any associated model parameters such as numbers and types of surface sites, surface area, relevant sorption reactions and their associated equilibrium constants are still debated. Generalized SCMs, such as the Dzombak and Morel surface complexation model, are developed on the basis of laboratory data on sorption of different ions to pure mineral phases. In natural systems, however, sorption is often controlled by an assemblage of multiple mineral phases. Two major alternative model approaches have emerged to account for the resulting, more complex adsorption behavior exhibited by natural sediments, i.e., the component additivity (CA) and generalized composite (GC) approaches (Arnold *et al.*, 2001; Davis *et al.*, 1998; 2005).

The CA approach assumes that the sorbent phase assemblage is composed of a mixture of identifiable and quantifiable sorbing minerals. In this case, surface chemical reactions are known for individual mineral components, e.g., from laboratory sorption experiments on goethite or amorphous Fe-oxides. Thus, depending on the estimate or measurement of the relative amounts of different sorbing minerals within the sorbent phase assemblage, sorption on each pure phase mineral is combined in an additive fashion according to the relative concentrations or surface areas of each mineral present in the natural substrate (Davis *et al.*, 1998; 2005). Adsorption by this mixture of mineral components can then be predicted in an equilibrium calculation, without requiring fitting of experimental data for the specific mixture (Payne *et al.*, 2013).

The GC approach acknowledges that the sorption properties of a natural sediment may be too complex to be quantified in terms of the contributions by individual mineral phases to adsorption and/or that the contribution of single mineral components to the macroscopic sorption behavior of the sediment may not simply be additive. Following the GC approach, a surface complexation model for natural sediment is established by fitting experimental sorption data for the mineral assemblage as a whole (Bond *et al.*, 2008; Hyun *et al.*, 2009; Payne *et al.*, 2004). This may not provide an accurate representation of surface complexation reactions at the molecular scale; however, the GC approach allows for an accurate description of the macroscopic sorption behavior of the natural system at hand.

Surface complexation models have been incorporated into subsurface transport models for several field-scale reactive transport modeling studies. Appelo *et al.* (2002) provided an excellent example of the usefulness of SCMs to simulate arsenic transport in Bangladesh groundwaters. They used a modified version of the Dzombak and Morel (1990) surface complexation model to illustrate mobilization of arsenic through desorption by bicarbonate. Also, Postma *et al.* (2007) investigated the mobilization of arsenic in a floodplain aquifer in Vietnam. They constructed a 1D reactive transport model with PHREEQC, which simulated to mobilization of arsenic through reductive dissolution of Fe-oxides, followed by subsequent re-adsorption on the surface of the remaining Fe-oxides. The employed database was compiled from different sources. Aqueous complexation reactions for arsenic were adopted from a compilation provided by Langmuir *et al.* (2006), while arsenic adsorption and desorption reactions were simulated with the database of Dzombak and Morel (1990), extended with additional sorption reactions for carbonate (Appelo *et al.*, 2002) and silica species (Swedlund and Webster, 1999). Another simulation example is a 1D PHREEQC reactive transport model that was developed by Charlet *et al.* (2007). It demonstrated the role of phosphate, bicarbonate and aqueous iron in mobilizing an As plume within a 3000-m aquifer section in West Bengal. A combination of surface complexation constants was adopted from Dzombak and Morel (1990) and Appelo *et al.* (2002) and the surface complexation capacity of the aquifer was coupled to the simulated amount of ferrihydrite. Similarly, Stollenwerk *et al.* (2007) successfully used PHREEQC for a 1D reactive transport modeling study to simulate the fate of arsenic. Again, adsorption reactions were modeled using mostly the surface complexation model of Dzombak and Morel (1990), while equilibrium constants for the surface complexation reactions for adsorption of arsenic, phosphate, silica and bicarbonate on aquifer sediments were fitted to experimental data. The model was used to demonstrate the capacity of oxidized sediments for remediating As-contaminated groundwater at a site near Dhaka, Bangladesh.

More recently a small number of multi-dimensional reactive transport modeling studies of the fate of arsenic have been reported. A 2D modeling study of As in groundwater discharging to Waquoit Bay, MA was undertaken by Jung *et al.* (2009). The performance of SCM *versus* empirical formulations in simulating the observed groundwater As distribution was evaluated using the simulation code PHT3D (Prommer *et al.*, 2003). Using the same code Wallis *et al.* (2010) developed a process-based description of the coupled physical and geochemical processes controlling the fate of arsenic during a deep well injection experiment in the Netherlands. The numerical model was further expanded to subsequently simulate the fate of arsenic in an aquifer storage and recovery (ASR) system in Florida (Wallis *et al.*, 2011). In the latter study the numerical modeling illustrated that pyrite oxidation and the precipitation/dissolution of amorphous iron-oxides (HFO) together with competitive displacement of As from sorption sites on HFO by competing anions were the key chemical processes that controlled the mobility of arsenic at the investigated aquifer storage and recovery (ASR) site. A detailed assessment of arsenic partitioning among mineral phases, surface complexes and aqueous phases during injection, storage and recovery was generated. Adsorption of arsenic was simulated using the diffuse double-layer model and arsenic-HFO surface complexation data was obtained from Dzombak and Morel (1990) and Appelo *et al.* (2002). In these models, the sorption capacity of the aquifer was linked to the amount of ferrihydrite.

2.3.4 *Mineral dissolution and precipitation*

Besides redox and sorption-desorption reactions, precipitation and dissolution of minerals are additional processes which can have a significant effect on the fate of arsenic. Dissolution reactions involve the erosion of the structure of minerals. Trace elements such as arsenic, uranium or lead, which can sometimes be present as major constituents of minerals or, more frequently, as impurities in various minerals can be released to groundwater as the host mineral dissolves. Prominent examples of arsenic release have been associated with the dissolution of sulfide minerals, most notably As-rich pyrite (FeS_2), which has been reported to contain up to 10 wt% arsenic (Price and Pichler, 2007), arsenopyrite ($FeAsS$), realgar (AsS) and orpiment (As_2S_3). These minerals are generally regarded as a primary source for As under oxidizing conditions (Plant *et al.*, 2007; Sracek *et al.*, 2004; Welch *et al.*, 2000), while, under reducing groundwater conditions, the reductive dissolution of Fe-oxides can be an important release mechanism for As (Burnol, 2007; Dixit and Hering, 2003; Smedley and Kinniburgh, 2002).

At the same time, (co)-precipitation can, sometimes very strongly, mitigate the mobilization of arsenic. For instance, the precipitation of As-bearing sulfides under reducing conditions, or the precipitation of ferrihydrite under oxidizing conditions may reduce dissolved arsenic concentrations through incorporation of arsenic into the mineral structure at it forms (O'Day *et al.*, 2004; Rittle *et al.*, 1995; Root *et al.*, 2009; Saunders *et al.*, 2008; Stollenwerk, 2003; Wolthers *et al.*, 2008). Precipitation-dissolution and adsorption-desorption reactions are therefore not independent and occur simultaneously.

In reactive transport models, dissolution-precipitation of minerals can be represented as equilibrium or kinetic reactions, or both. Strictly speaking, mineral equilibrium should only be invoked in cases where the local equilibrium assumption (LEA) applies, that is, where the timescale of the precipitation-dissolution reaction is relatively fast compared to the transport timescale. Where this is not the case, i.e., where mineral reactions are slow and mineral equilibria are not attained within the time that corresponds to a travel-distance equivalent to the grid cell size imposed by model discretization, the reactions should be simulated as kinetically controlled reactions. Reaction rate equations exist for many common mineral dissolution and/or precipitation reactions that occur in typical groundwater systems. These rate formulations can be incorporated into geochemical and reactive transport models such as PHREEQC (Parkhurst and Appelo, 1999) and associated models such as PHAST (Parkhurst *et al.*, 2010) and PHT3D (Prommer *et al.*, 2003) to simulate kinetically controlled precipitation/dissolution reactions. Similarly kinetic source terms for arsenic release or uptake (e.g., Lasaga, 1998; Matsunaga *et al.*, 1993; Schreiber and Rimstidt, 2013) can also be formulated. Where more detailed reaction rate studies are unavailable, a commonly applied reaction rate formulation for dissolution and precipitation of minerals is (e.g., Lasaga, 1998):

$$R_k = k_k \left[1 - \left(\frac{IAP}{K_{SP}} \right) \right] \tag{2.7}$$

where k_k is an effective reaction rate constant and IAP/K_{SP} is the saturation ratio, i.e., the ratio of the ion activity product and the solubility product constant for the mineral.

The study by Wallis *et al.* (2010) is an example where kinetically controlled dissolution reactions were used to define an arsenic source. This study presented simulations of an aquifer storage, transfer and recovery (ASTR) operation in the Netherlands where oxygenated water was injected into a reducing, pyritic aquifer. During injection, arsenic associated with pyrite, was found to become unstable under the progressively more oxidizing conditions and thereby released into the aqueous solution. Reactions for pyrite oxidation by oxygen and nitrate were therefore included in the reactive transport model, based on previously proposed and applied rate expressions (Eckert and Appelo, 2002; Prommer and Stuyfzand, 2005; Williamson and Rimstidt, 1994):

$$r_{pyr} = \left[(C_{O_2}^{0.5} + f_2 C_{NO_3^-}^{0.5}) C_{H^+}^{-0.11} \left(10^{-10.19} \frac{A_{pyr}}{V} \right) \left(\frac{C}{C_0} \right)_{pyr}^{0.67} \right] \tag{2.8}$$

where r_{pyr} is the specific oxidation rate for pyrite, C_{O_2}, $C_{NO_3}^-$, and C_H^+ are the oxygen, nitrate and proton concentrations in groundwater, A_{pyr}/V is the ratio of mineral surface area to solution volume and (C/C_0) is a factor that accounts for changes in A_{pyr} as pyrite is progressively depleted by the reaction. *Aqua regia* extraction of sediment cores found As, Co, Ni and Zn to be associated with pyrite with a most probable stoichiometry of $Fe_{0.98}Co_{0.0037}Ni_{0.01}Zn_{0.01}S_2As_{0.0053}$ at the site. Arsenic release from pyrite was simulated by kinetic dissolution of an As-bearing pyrite component that was stoichiometrically linked to the pyrite oxidation rate (at a molar ratio of 0.0053).

Postma *et al.* (2007) used an alternative approach to quantify the mobilization of arsenic in a shallow Holocene aquifer on the Red River flood plain near Hanoi, Vietnam. In their case organic carbon decomposition in the anoxic aquifer induced distinct redox zones that were dominated by the reduction of Fe-oxides and methanogenesis. In their 1D PHREEQC model As(V) was incorporated as a minor constituent at an As/Fe molar ratio of 0.0025 within the Fe-oxides originally present in the aquifer. Consequently the progressive reductive dissolution of Fe-oxide triggered As(V) release into the aqueous phase in the model.

2.4 SUMMARY AND OUTLOOK

In this chapter a general overview on numerical modeling approaches, common tools and some model applications was provided. From this overview it is evident that enormous progress has been made over the last decade, and both understanding of processes as well as modeling tools have now matured to the degree that geochemical and reactive transport modeling can routinely be part of studies that assess and quantify the fate of arsenic in groundwater. To date, most model applications have been focused on using modeling as a tool to analyze existing laboratory and field data whereby the calibration of models to measured data was generally performed via trial and error. Over the next decade further enhanced computational resources such as high-performance computing clusters and cloud computing will see, similar to trends in groundwater flow modeling, a shift to an increased use of automatic parameter estimation tools that allow a more rigorous estimation of optimal model parameters and evaluation of their sensitivities. Within such frameworks model predictions that forecast the long-term fate of arsenic will be accompanied by advanced assessments of model uncertainty.

REFERENCES

Allison, J.D., Brown, D.S. & Novagradac, K.J.: MINTEQA2/PRODEFA2, A geochemical assessment model for environmental systems: version 3.0. US Environmental Protection Agency, Athens, GA, 1990.

Anderson, M.P. & Woessner, W.W.: *Applied groundwater modeling simulation of flow and advective transport*. Academic Press, San Diego, CA, 1992.

Appelo, C.A.J. & Postma, D.: *Geochemistry, groundwater and pollution*. A.A. Balkema Publishers, Amsterdam, The Netherlands, 2005.

Appelo, C.A.J., Van Der Weiden, M.J.J., Tournassat, C. & Charlet, L.: Surface complexation of ferrous iron and carbonate on ferrihydrite and the mobilization of arsenic. *Environ. Sci. Technol.* 36:14 (2002), pp. 3096–3103.

Appelo, C.A.J. & De Vet, W.W.J.M.: Modeling in situ iron removal from groundwater with trace elements such as As. In: A.H. Welch & K.G. Stollenwerk (eds): *Arsenic in groundwater*. Kluwer Academic, Boston, MA, 2003, pp. 381–401.

Barry, D.A., Prommer, H., Miller, C.T., Engesgaard, P., Brun, A. & Zheng, C.: Modeling the fate of oxidisable organic contaminants in groundwater. *Adv. Water Resour.* 25 (2002), pp. 945–983.

Bear, J.: *Dynamics of fluids in porous media*. Dover Publications Inc., New York, NY, 1972.

Bear, J. & Verruijt, A.: Modeling groundwater flow and pollution. D. Reidel Publishing Co., Dordrecht, The Netherlands, 1987.

Baedecker, M.J., Cozarelli, I.M., Siegel, D.I., Bennett, P.C. & Eganhouse, R.P.: Crude oil in a shallow sand and gravel aquifer. 3. Biogeochemical reactions and mass balance modeling. *Appl. Geochem.* 8 (1993), pp. 569–586.

Brun, A. & Engesgaard, P.: Modeling of transport and biogeochemical processes in pollution plumes: literature review and model development. *J. Hydrol.* 256 (2002), pp. 211–227.

Burnol, A., Garrido, F., Baranger, P., Joulian, C., Dictor, M.-C., Bodénan, F., Morin, G. & Charlet, L.: Decoupling of arsenic and iron release from ferrihydrite suspension under reducing conditions: a biogeochemical model. *Geochem. Trans.* 8:12 (2007).

Champ, D.R., Gulens, J. & Jackson, R.E.: Oxidation-reduction sequences in groundwater flow systems. *Can. J. Earth Sci.* 16 (1979), pp. 12–23.

Chapelle, F.H. & Lovley, D.R.: competitive exclusion of sulfate-reduction by Fe(III)-reducing bacteria: a mechanism for producing discrete zones of high-iron ground water. *Ground Water* 30 (1992), pp. 29–36.

Charlet, L., Chakraborty, S., Appelo, C.A.J., Roman-Ross, G., Nath, B., Ansari, A.A., Lanson, M., Chatterjee, D. & Mallik, S.B.: Chemodynamics of an arsenic "hotspot" in a West Bengal Aquifer: a field and reactive transport modeling study. *Appl. Geochem.* 22:7 (2007), pp. 1273–1292.

Chiang, W.-H. & Kinzelbach, W.: *3D-groundwater modeling with PMWIN: A simulation system for modeling groundwater flow and pollution*. Springer, Heidelberg, Germany, 2000.

Christensen, T.H., Bjerg, P.L., Banwart, S.A., Jakobsen, R., Heron, G. & Albrechtsen, H.-J.: Characterization of redox conditions in groundwater contaminant plumes. *J. Cont. Hydrol.* 45 (2000), pp. 165–241.

Clement, T.P.: RT3D – A modular computer code for simulating reactive multi-species transport in 3-dimensional groundwater aquifers. Pacific Northwest National Laboratory, Richland, WA, 1997.

Darland, E.J. & Inskeep, P.W.: Effects of pore water velocity on transport of arsenate. *Environ. Sci. Technol.* 31:3 (1997), pp. 704–709.

Davis, J.A. & Kent, D.B.: Surface complexation modeling in aqueous geochemistry. Chapter 5 in: M.F., Jr. Hochella, A.F. White & P.H. Ribbe (eds): *Reviews in mineralogy: mineral-water interface geochemistry*, Vol. 23. Mineralogical Society of America, Washington DC, 1990, pp. 177–248.

Davis, J.A. & Leckie. J.O.: Surface ionization and complexation at the oxide/water interface. II. Surface properties of amorphous iron oxyhydroxide and adsorption of metal ions. *J. Colloid Interface Sci.* 67 (1978), pp. 90–107.

Decker, D.L., Simunek, J., Tyler, S.W., Papelis, C. & Logsdon, M.J.: Variably saturated reactive transport of arsenic in heap-leach facilities. *Vadose Zone J.* 5:1 (2006), pp. 430–444.

Diersch, H.-J.: Interactive, graphics-based finite-element simulation system FEFLOW for modelling groundwater flow, contaminant mass and heat transport processes. User's manual version 4.6. WASY. Institute for Water Resources Planning and System Research Ltd., Berlin, Germany, 1997.

Dixit, S. & Hering, J.G.: Comparison of arsenic(V) and arsenic(III) sorption onto iron oxide minerals: implications for arsenic mobility. *Environ. Sci. Technol.* 37:18 (2003), pp. 4182–4189.

Domenico, P.A. & Schwartz, W.: *Physical and chemical hydrogeology*. Wiley, New York, NY, 1998.

DPHE, BGS & MML: Groundwater studies for arsenic contamination in Bangladesh. Phase I: rapid investigation phase. Main report and five supplementary volumes. Dep. of Public Health Engineering, Gov. of Bangladesh, BGS and Mott Macdonald Ltd (UK), 1999.

Dzombak, D.A. & Morel, F.M.M.: *Surface complexation modeling hydrous ferric oxide*. John Wiley & Sons Inc, New York, NY, 1990.

Eckert, P. & Appelo, C.A.J.: Hydrogeochemical modeling of enhanced benzene, toluene, ethylbenzene, xylene (BTEX) remediation with nitrate. *Water Resour. Res.* 38:8 (2002), pp. 5-1–5-11.

Fetter, C.W.: *Applied hydrogeology*. Prentice-Hall, Upper Saddle River, NJ, 2001.

Fiori, A. & Dagan, G.: Concentration fluctuations in transport by groundwater: comparison between theory and field experiments. *Water Resour. Res.* 35:1 (1999), pp. 105–112.

Freundlich, H.: *Colloid and capillary chemistry*. Methuen, London, UK, 1926.

Gao, H., Butler, A., Wheater, H. & Vesovic, V.: Chemically reactive multicomponent transport simulation in soil and groundwater: 1. Model development and evaluation. *Environ. Geol.* 41:3–4 (2001), pp. 274–279.

Goldberg, S.: Adsorption of arsenate and arsenite on oxides and clay minerals. *Soil Sci. Soc. Am. J.* 66 (2002), pp. 413–421.

Goldberg, S.R. & Criscenti, L.J.: Modeling adsorption of heavy metals and metalloids by soil components. In: A. Violante, P.M. Huang & Stotzsky, G. (eds): *Biophysico-chemical processes of heavy metals and metalloids in soil environments*. J. Wiley & Sons, New York, NY, 2007, pp. 215–264.

Goldberg, S.R., Criscenti, L.J., Turner, D.R., Davis, J.A. & Cantrell, K.J.: Adsorption-desorption processes in subsurface reactive transport modeling. *Vadose Zone J.* 6:3 (2007), pp. 407–435.

Greskowiak, J., Prommer, H., Vanderzalm, J., Pavelic, P. & Dillon, P.: Modeling of carbon cycling and biogeochemical changes during injection and recovery of reclaimed water at Bolivar, South Australia. *Water Resour. Res.* 41:10 (2005), W10418.

Harbaugh, A.W.: Modflow-2005, the US Geological Survey modular ground-water model—the ground-water flow process. US Geological Survey Techniques and Methods 6-A16. Reston, VA, USA, 2005.

Harbaugh, A.W., Banta, E.R., Hill, M.C. & McDonald, M.G.: Modflow-2000, the US Geological Survey modular ground-water model: user guide to modularization concepts and the ground-water flow process. US Geological Survey Open-File Report 00-92. Reston, VA, USA, 2000.

Hiemstra, T. & Van Riemsdijk, W.A.: Surface structural approach to ion adsorption: the charge distribution (Cd) model. *J. Colloid Interface Sci.* 179 (1996), pp. 488–508.

Hsia, T.H., Lo, S.L. & Lin, C.F.: As(V) adsorption on amorphous iron oxide: triple layer modelling. *Chemosphere* 25:12 (2002), pp. 1825–1837.

Hunter, K.S., Wang, Y.F. & Van Cappellen, P.: Kinetic modeling of microbially-driven redox chemistry of subsurface environments: coupling transport, microbial metabolism and geochemistry. *J. Hydrol.* 209: 1–4 (1998), pp. 53–80.

Hyun, S.P., Fox, P.M., Davis, J.A., Campbell, K.M., Hayes, K.F. & Long, P.E.: Surface complexation modeling of U(VI) adsorption by aquifer sediments from a former mill tailings site at Rifle, Colorado. *Environ. Sci. Technol.* 43:24 (2009), pp. 9368–9373.

Jensen, K.H., Bitsch, K. & Bjerg, P.L.: Large-scale dispersion experiments in a sandy aquifer in Denmark: observed tracer movements and numerical analyses. *Water Resour. Res.* 29:3 (1993), pp. 673–696.

Jeppu, G.P., Clement, T.P., Barnett, M.O. & Lee, K.: A modified batch reactor system to study equilibrium-reactive transport problems. *J. Contam. Hydrol.* 129–130, pp. 2–9.

Kirk, M.F., Holm, T.R., Park, J., Jin, Q., Sanford, R.A., Fouke, B.W. & Bethke, C.M.: Bacterial sulfate reduction limits natural arsenic contamination in groundwater. *Geology* 32:11 (2004), pp. 953–956.

Konikow, L.F.: The secret to successful solute-transport modeling. *Ground Water* 49:2 (2012), pp. 144–159.

Langmuir, D.: The adsorption of gases on plane surfaces of glass, mica, and platinum. *J. Am. Chem. Soc.* 40 (1918), pp. 1361–1403.

Langmuir, D.L., Mahoney, J. & Rowson, J.: Solubility products of amorphous ferric arsenate and crystalline scorodite (FeAsO$_4$·2H$_2$O) and their application to arsenic behavior in buried mine tailings. *Geochim. Cosmochim. Acta* 70:12 (2006), pp. 2942–2956.

Lasaga, A.C.: Kinetic theory in earth sciences. Princeton University Press, Princeton, NY, 1998.

Levine, A.D., Libelo, E.L., Bugna, G., Shelley, T., Mayfield, H. & Stauffer, T.B.: Biogeochemical assessment of natural attenuation of JP-4-contaminated ground water in the presence of fluorinated surfactants. *Sci. Total Environ.* 208:3 (1997), pp. 179–195.

Lichtner, P.C., Steefel, C.I. & Oelkers, E.H. (eds): Reactive transport in porous media. *Reviews in Minerology* 34. Mineralogical Society of America, 1996.

MacQuarrie, K.T.B. & Sudicky, E.A.: Multicomponent simulation of wastewater-derived nitrogen and carbon in shallow unconfined aquifers: i. model formulation and performance. *J. Contam. Hydrol.* 47:1 (2001), pp. 53–84.

Massmann, G., Pekdeger, A. & Merz, C.: Redox processes in the Oderbruch Polder groundwater flow system in Germany. *Appl. Geoch.* 19:6 (2004), pp. 863–886.

Masue, Y., Loeppert, R.H. & Kramer, T.A.: Arsenate and arsenite adsorption and desorption behavior on coprecipitated aluminum: iron hydroxies. *Environ. Sci. Technol.* 41 (2007), pp. 837–842.

Matsunaga, T., Karametaxas, G., von Gunten, H.R. & Lichtner, P.C.: Redox chemistry of iron and manganese minerals in river-recharged aquifers: a model interpretation of a column experiment. *Geochim. Cosmochim. Acta* 57 (1993), pp. 1691–1704.

McMahon, P.B. & Chapelle, F.H.: Microbial production of organics acids in aquitard sediments and its role in aquifer geochemistry. *Nature* 349 (1991), pp. 233–235.

Nath, B., Chakraborty, S., Burnol, A., Stüben, D., Chatterjee, D. & Charlet, L.: Mobility of arsenic in the sub-surface environment: an integrated hydrogeochemical study and sorption model of the sandy aquifer materials. *J. Hydrol.* 364 (2009,) pp. 236–248.

O'Day, P.A., Vlassopoulos, D., Root, R. & Rivera, N.: The influence of sulfur and iron on dissolved arsenic concentrations in the shallow subsurface under changing redox conditions. *PNAS* 101 (2004), pp. 13,703–13,708.

Park, J., Sanford, R.A. & Bethke, C.M.: Microbial activity and chemical weathering in the Middendorf Aquifer, South Carolina. *Chem. Geol.* 258:3–4 (2009), pp. 232–241.

Parkhurst, D. & Appelo, C.: User's guide to PHREEQC (Version 2) – A computer program for speciation, batch-reaction, one-dimensional transport, and inverse geochemical calculations. US Geological Survey, Water-Resources Investigations Report 99-4259, 1999.

Parkhurst, D.L., Thorstenson, D.C. & Plummer, L.N.: PHREEQE: A computer program for geochemical calculations. US Geological Survey, Water Resources Division, Water-Resources Investigations Report 80-96, V, 1980.

Parkhurst, D., Christenson, S. & Breit, G.N.: Ground-water-quality assessment of the Central Oklahoma Aquifer, Oklahoma-geochemical and geohydrological investigations. US Geological Survey, Water-Supply Paper 2357 Guidelines for Thermodynamic Sorption Modelling in the Context of Radioactive Waste Disposal, 1996.

Parkhurst, D.L., Kipp, K.L., Engesgaard, P. & Charlton, S.R.: PHAST – A program for simulating ground-water flow, solute transport, and multicomponent geochemical reactions. US Geological Survey, Techniques and Methods 6-A8, 2004.

Parkhurst, D., Kipp, K.L. & Charlton, S.R.: PHAST version 2 – A program for simulating groundwater flow, solute transport, and multicomponent geochemical reactions: US Geological Survey, Techniques and Methods, 6–A35, 2010.

Payne, T.E., Davis, J.A., Lumpkin, G.R., Chisari, R. & Waite, T.D.: Surface complexation model of uranyl sorption on Georgia kaolinite. *Appl. Clay Sci.* 26: 1–4 (2004), pp. 151–162.

Payne, T.E., Brendler, V., Ochs, M., Baeyens, B., Brown, P.L., Davis, J.A., Ekberg, C., Kulik, D.A., Lutzenkirchen, J., Missana, T., Tachi, Y., Van Loon, L.R. & Altmann, S.: Guidelines for thermodynamic sorption modelling in the context of radioactive waste disposal. *Environ. Modell. Softw.* 42 (2013), pp. 143–156.

Plant, J.A., Kinniburgh, D.G., Smedley, P.L., Fordyce, F.M., Klinck, B.A., Heinrich, D.H. & Karl, K.T.: Arsenic and selenium. In: *Treatise on geochemistry*, pp. 17–66. Pergamon, Oxford, UK, 2007.

Plummer, L.N., Parkhurst, D.L. & Thorstenson, D.C.: Development of reaction models in ground-water systems. *Geochim. Cosmochim. Acta* 47 (1983), pp. 665–686.

Postma, D. & Jakobsen, R.: Redox zonation: equilibrium constraints on the Fe (III)/SO_4-reduction Interface. *Geochim. Cosmochim. Acta* 60:17 (1996), pp. 3169–3175.

Postma, D., Larsen, F., Minh Hue, N.T., Duc, M.T., Viet, P.H., Nhan, P.Q. & Jessen, S.: Arsenic in groundwater of the Red River Floodplain, Vietnam: controlling geochemical processes and reactive transport modeling. *Geochim. Cosmochim. Acta* 71:21 (2007), pp. 5054–5071.

Price, R.E. & Pichler, T.: Abundance and mineralogical association of arsenic in the Suwannee Limestone (Florida): implications for arsenic release during water-rock interaction. *Chem. Geol.* 228:1–3 (2006), pp. 44–56.

Prommer, H. & Stuyfzand, P.J.: Identification of temperature-dependent water quality changes during a deep well injection experiment in a pyritic aquifer. *Environ. Sci. Technol.* 39:7 (2007), pp. 2200–2209.

Prommer, H., Barry, D.A. & Davis, G.B.: Modelling of physical and reactive processes during biodegradation of a hydrocarbon plume under transient groundwater flow conditions. *J. Contam. Hydrol.* 59:1–2 (2002), pp. 113–131.

Prommer, H., Barry, D.A. & Zheng, C.: MODFLOW/MT3DMS-based reactive multicomponent transport modeling. *Ground Water* 41:2 (2003), pp. 247–257.

Prommer, H., Tuxen, N. & Bjerg, P.L.: Fringe-controlled natural attenuation of phenoxy acids in a landfill plume: integration of field-scale processes by reactive transport modeling. *Environ. Sci. Technol.* 40 (2006), pp. 4732–4738.

Rittle, K.A., Drever, J.I. & Colberg, P.J.S.: Precipitation of arsenic during bacterial sulfate reduction. *Geomicrobiol. J.* 13:1(1995), pp. 1–11.

Root, R.A, Vlassopoulos, D., Rivera, N.A., Rafferty, M.T., Andrews, C. & O'Day, P.A.: Speciation and natural attenuation of arsenic and iron in a tidally influenced shallow aquifer. *Geochim. Cosmochim. Acta* 73 (2009), pp. 5528–5553.

Saunders, J.A., Lee, M.-K., Shamsudduha, M., Dhakal, P., Uddin, A., Chowdury M.T. & Ahmed, K.M.: Geochemistry and mineralogy of arsenic in (natural) anaerobic groundwaters. *Appl. Geochem.* 23 (2008), pp. 3205–3214.

Schreiber, M.E. & Rimstidt, J.D.: Trace element source terms for mineral dissolution. *Appl. Geochem.* 37 (2013), pp. 94–101.

Šimůnek, J. & Suarez, D.L.: Two-dimensional transport model for variably saturated porous media with major ion chemistry. *Water Resour. Res.* 30 (1994), pp. 1115–1133.

Šimůnek, J., Suarez, D.L. & Sejna, M.: The UNSATCHEM software package for simulating the one-dimensional variably saturated water flow, heat transport, carbon dioxide production and transport, and multicomponent solute transport with major ion equilibrium and kinetic chemistry. US Salinity Laboratory, Research Report No. 141, Riverside, CA, USA, 1996.

Smedley, P.L. & Kinniburgh, D.G.: A review of the source, behavior and distribution of arsenic in natural waters. *Appl. Geochem.* 17:5 (2002), pp. 517–568.

Sracek, O., Bhattacharya, P., Jacks, G., Gustafsson, J.-P. & Brömssen, M.V.: Behavior of arsenic and geochemical modeling of arsenic enrichment in aqueous environments. *Appl. Geoch.* 19:2 (2004), pp. 169–180.

Stollenwerk, K.G.: Geochemical processes controlling transport of arsenic. In: A.H. Welch & K.G. Stollenwerk (eds): *Arsenic in groundwater*. Kluwer, Boston, MA, 2003, pp. 67–100.

Stollenwerk, K.G., Breit, G.N., Welch, A.H., Yount, J.C., Whitney, J.W., Foster, A.L., Uddin, M.N., Majumder, R.K. & Ahmed, N.: Arsenic attenuation by oxidized aquifer sediments in Bangladesh. *Sci. Total Environ.* 379:2–3 (2007), pp. 133–150.

Stuyfzand, P.J.: *Hydrochemistry and hydrology of the coastal dune area of the Western Netherlands*. PhD Thesis, Vrije University Amsterdam, Amsterdam, The Netherlands, 1993.

Stuyfzand, P.J. & Timmer, H.: Deep well injection at the Langerak and Nieuwegein sites in the Netherlands: chemical reactions and their modelling. KIWA-SWE 96.006, Nieuwegein, The Netherlands, 1999.

Stumm, W. & Morgan, J.J.: *Aquatic chemistry*. John Wiley – Interscience, New York, NY, 1996.

Stumm, W., Kummert, R. & Sigg, L.: A ligand exchange model for the adsorption of inorganic and organic ligands at hydrous oxide interfaces. *Croat. Chem. Acta* 53 (1980), pp. 291–312.

Swedlund, P.J. & Webster, J.G.: Adsorption and polymerization of silicic acid on ferrihydrite and its effect on arsenic adsorption. *Water Res.* 33 (1999), pp. 3413–3422.

Wang, H. & Anderson, M.: *Introduction to groundwater modeling: finite difference and finite element methods*. Academic Press, San Diego, CA, 1982.

Wallis, I., Prommer, H., Simmons, C.T., Post, V. & Stuyfzand, P.J.: Evaluation of conceptual and numerical models for arsenic mobilization and attenuation during managed aquifer recharge. *Environ. Sci. Technol.* 44:13 (2010), pp. 5035–5041.

Wallis, I., Prommer, H., Pichler, T., Post, V., Norton, S.B., Annable, M.B. & Simmons, C.T.: Process-based reactive transport model to quantify arsenic mobility during aquifer storage and recovery of potable water. *Environ. Sci. Technol.* 45:16 (2011), pp. 6924–6931.

Welch, A.H., Westjohn, D.B., Helsel, D.R. & Wanty, R.B.: Arsenic in ground water of the United States: occurrence and geochemistry. *Ground Water* 38:4 (2000), pp. 589–604.

Williamson, M.A. & Rimstidt, J.D.: The kinetics and electrochemical rate-determining step of aqueous pyrite oxidation. *Geochim. Cosmochim. Acta* 58:24 (1994), pp. 5443–5454.

Wolthers, M., Charlet, L., Van Der Weijden, C.H., Van Der Linde, P.R. & Rickard, D.: Arsenic mobility in the ambient sulfidic environment: sorption of arsenic(V) and arsenic(III) onto disordered mackinawite. *Geochim. Cosmochim. Acta* 69:14 (2005), pp. 3483–3492.

Yeh, G.T., Sharp-Hansen, S., Lester, B., Strobl R. & Scarbrough, J.: 3DFEMWATER/3DLEWASTE: Numerical codes for delineating wellhead protection areas in agricultural regions based on the assimilative capacity. US Environmental Protection Agency, Athens, GA, USA, 1992.

Zhang, Y.C., Slomp, C.P., Broers, H.P., Passier, H.F. & Van Cappellen, P.: Denitrification coupled to pyrite oxidation and changes in groundwater quality in a shallow sandy aquifer. *Geochim. Cosmochim. Acta* 73:22 (2009), pp. 6716–6726.

Zheng, C. & Wang, P.P.: MT3DMS, A modular three-dimensional multi-species transport model for simulation of advection, dispersion and chemical reactions of contaminants in groundwater systems; documentation and user's guide. US Army Engineer Research and Development Center Contract Report 202, 1999.

CHAPTER 3

Phytostabilization of arsenic

Claes Bergqvist & Maria Greger

3.1 INTRODUCTION

Pollutants from natural and anthropogenic sources can contaminate areas, having severe effects on natural ecosystems and human land use. Sites contaminated due to military or industrial activities, including wood impregnation, oil refining, gas or coal processing, and ammunition and pesticide production, are often contaminated with arsenic (As) and in need of restoration. In certain sedimentary rocks, such as shales, with a high organic content, As levels can be elevated. Certain areas that contain high levels of toxic pollutants including As are so extensive that other forms of remediation than stabilization are unrealistic. At such sites, natural vegetative growth is often scarce due to high pollutant and/or low nutrient levels. Unvegetated areas may cause problems for both human and animal populations, for example, due to inhalation of contaminated airborne particles or through food chains (Mench *et al.*, 2000).

Phytotechnologies may offer a cost-effective alternative to established remediation techniques for the restoration of contaminated areas (Mench *et al.*, 2010). Phytostabilization can be defined as a method to immobilize pollutants using plants and plant-associated microbes (Cunningham *et al.*, 1995; Ward and Singh, 2004). The ultimate goal of phytostabilization is to create a self-sustaining, vegetative cap for the long-term control of polluting agents, reducing the availability of pollutants by preventing dispersal by wind and water. Phytostabilization has advantages over conventional mechanical remediation technologies, including being economically feasible and less destructive. The primary focus of phytostabilization is to prevent the pollutants from translocating to aboveground plant tissues and to sequester the pollutants in the rhizosphere through precipitation, i.e., accumulated by the plant roots and immobilized onto soil particles (Mendez and Maier, 2008a) (Fig. 3.1).

Successful phytostabilization of heavy metals such as Cd, Pb, Zn, and B has been reported in diverse materials, such as mine tailings and wood wastes (Brown *et al.*, 2005; Robinson *et al.*, 2007; Stoltz and Greger, 2002). In this chapter, we review current knowledge in the area of As phytostabilization.

3.2 ARSENIC

Arsenic is present as a ubiquitous trace element. Several forms or species of As exist in soil and water (Fig. 3.2). Arsenic is carcinogenic and its toxicity is dependent on speciation. The inorganic species of As are regarded as more harmful than the organic species to a wide range of living organisms, with arsenite being more toxic than arsenate (Meharg and Hartley-Whitaker, 2002). The toxicity of organic As compounds such as arsenobetaine, found, for example, in fish, is generally considered low (Kaize *et al.*, 1985). However, trivalent methylated intermediates, such as methylarsonous acid and dimethylarsinous acid produced in the metabolic detoxification pathways of As in the human liver, are more toxic than inorganic As species (Dopp *et al.*, 2010). The major route for As intake by humans is through drinking water. Based on the risk of developing cancer due to lifetime As exposure, most countries have set the limit for As in drinking water

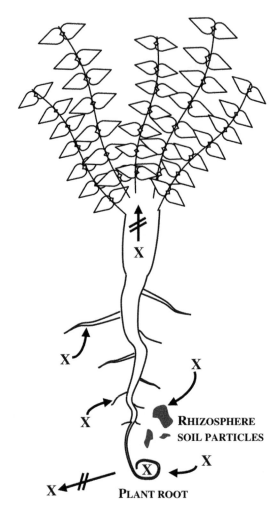

Figure 3.1.   Phytostabilization of pollutants (X). Accumulation in plant roots and precipitation onto soil particles in the rhizosphere immobilizes the pollutants, preventing translocation to the shoots and leaching to the surrounding surface and groundwater.

to $10\,\mu g\,L^{-1}$ (Commission Directive 2003/40/EC). No worldwide standards for As in food are currently available. Instead, the European Food Safety Authority (EFSA) has established an inorganic As exposure dosage of 0.3–$8\,\mu g\,As\,kg^{-1}$ bodyweight per day (EFSA, 2009).

A variety of environmental factors (e.g., pH, redox potential, adsorption reactions, and biological activity) influence the concentrations and speciation of soluble As in soils (Bhumbla and Keefer, 1994). The most common transformation pathways of As in soils are illustrated in Figure 3.3. With increased biological activity, the amount of organic As species increases. For example, approximately two thirds of the As in a water sample from the Salton Sea in California was in the form of dimethylarsinic acid due to the activity of phytoplankton (Andreae and Klumpp, 1979).

Arsenate and arsenite are the dominant As species in soils and waters. The redox potential is the most influential factor in arsenate–arsenite speciation under normal pH ranges in soil. With increasing redox potential, As transformation shifts toward an increased amount of arsenate ($pE > 10$), while arsenite dominates under low redox potential conditions ($pE < 6$) (Sadiq, 1997).

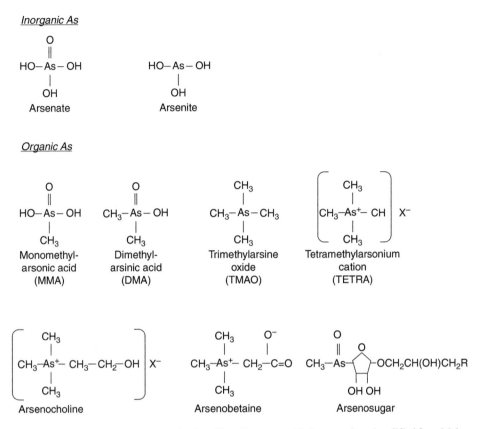

Figure 3.2. The most common As species found in soil, water, and living organisms (modified from Meharg and Hartley-Whitaker, 2002); X = accompanying anion.

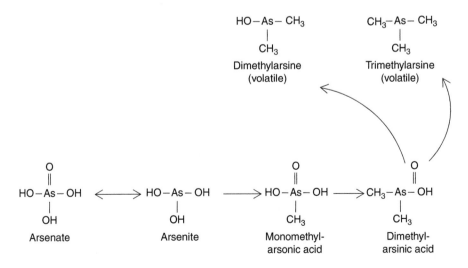

Figure 3.3. Transformations of As species in soils (modified from Wood, 1974, and Bhumbla and Keefer, 1994).

However, it is difficult to determine with any certainty what factors will influence the arsenate–arsenite speciation in soil. For example, in 40 soils treated similarly with regard to redox potential, the amount of arsenite and the arsenite/arsenate ratio varied considerably between the soils, indicating that a range of both abiotic and biotic processes influences the arsenate–arsenite speciation (Ackermann *et al.*, 2010).

The pH may influence As availability. Generally, soil pH > 7 increases the availability of anions such as arsenate and arsenite, while soil pH < 7 may favor the retention of anionic As due to an abundance of positive charges that neutralize the negative exchange sites in soil (Moreno-Jiménez *et al.*, 2012). However, basic pH may favor co-precipitation of As with, for example, sulfate or calcium, leading to reduced availability, and in pH below 2.5, arsenate may become fully protonated leading to increased availability (Moreno-Jiménez *et al.*, 2012). The availability also differs between arsenate and arsenite. For example, adsorption of arsenate to ferrihydrite in the pH range 4–9 decreased with increasing pH, while the opposite occurred for arsenite (Raven *et al.*, 1998). The stronger adsorption of arsenite with increasing pH was likely due to the formation of ferric arsenite.

Small amounts of the organic As species, such as monomethylarsonic acid (MMA) and dimethylarsinic acid (DMA), are also commonly present in soil (Bowell, 1994) (Fig. 3.3). The organic As species result mainly from the biological activity of microorganisms, primarily bacteria but also fungi (Wood, 1974). Under reduced conditions, bacteria can convert some of the organic As species into the volatile and highly toxic As species dimethylarsine (McBride and Wolfe, 1971) and trimethylarsine (Sadiq, 1997) (Fig. 3.3). For example, up to 240 mg of volatile As, mainly in the form of trimethylarsine, per hectare per year was released from a low-As-concentration paddy soil (containing 11 mg As kg^{-1} DW), leading to an estimated contribution from soils of 0.9–2.6% of the total global atmospheric input of As (Mestrot *et al.*, 2011).

3.3   SOIL COMPOSITION AND ARSENIC AVAILABILITY

The total concentration of As in soils is not the only determinant of its toxicity to living organisms. Instead, the availability and speciation of As are the most important factors determining its toxicity. For example, a five-fold higher toxicity of inorganic As has been observed in sandy than clayey soils (Sheppard, 1992). Cao *et al.* (2003) compared two soils containing similar amounts of total As: one was collected from an old abandoned wood preservation site and the other was spiked with As by the authors. The authors found that As was approximately five times more available in the laboratory-spiked soil than in the soil from the field. In the spiked soil, the As was predominately bound to labile aluminum oxides, whereas in the soil from the wood preservation site, As was bound to stable calcium-As and iron-As fractions, leading to the difference in As availability (Cao *et al.*, 2003). Ma *et al.* (2006) referred to the process of reduced availability of As in soil over time, as in the laboratory-spiked *versus* field soil case, as aging.

The geochemical form of As in soil strongly influences the soil As. Arsenopyrite has a low availability, whereas As trioxide has a high availability. Depending on the pentavalent As mineral present in the soil, As availability might vary between 0 and 41% (Meunier *et al.*, 2011). Iron strongly influences the availability of As through the formation of amorphous iron(III) arsenates and by replacing the surface hydroxyl groups of iron oxides with As (Kumpiene *et al.*, 2008). The availability of arsenate is usually low in soils due to the strong adsorption of arsenate to iron and aluminum oxides (Zhao *et al.*, 2010). Arsenite is more weakly adsorbed to soil particles and is therefore easily released into the soil solution or displays higher availability under low redox conditions (pE < 6). Iron-As oxides are also dissolved in soils with low redox potential (pE < 6) with a concomitant redox-induced transformation of As to arsenite, leading to an increased availability of arsenite (Zhao *et al.*, 2010).

Due to its anionic nature, complexation between As and negatively charged organic soil colloids is low, and the availability of As is therefore mainly dependent on adsorption to clay minerals, calcium carbonates, or iron or aluminum oxides or hydroxides (Sadiq, 1997). Organic matter increases the mobility of As due to physical competition and electrostatic interactions with soil particles, and through the formation of aqueous complexes containing As (Wang and Mulligan, 2009). Arsenic availability in soils with an organic matter content of more than 25% increased with time due to an original high content of available As that became mobile during mineralization (Meunier *et al.*, 2011). In soils with an organic matter content below 8%, aging did not influence the availability of As due to the presence of arsenopyrite and pentavalent As forms, which were not affected by organic matter (Meunier *et al.*, 2011). However, ternary complexes between ferric complexes of humic substances and arsenate accounted for 25–70% of the total As in a laboratory solution composed of humic acids, iron(III), and arsenate (Mikutta and Kretzschmar, 2011). The complexation between humic substances, iron, and As with a concomitant reduced availability of As, may lead to a revision of the general opinion of the relationship between organic matter and As in the near future. In addition, phosphate increases As mobility. In the normal pH range in soil (pH 4–8), arsenate exists mainly in its deprotonated form, as anions. In its anionic form, arsenate acts as a physical analogue to phosphate, leading to competition for adsorption sites on iron oxides/hydroxides in the soil (Zhao *et al.*, 2010). At an old chemical waste site, As mobility was correlated with the content of both organic matter and phosphate (Hartley *et al.*, 2009).

3.4   PLANT TRAITS IN PHYTOSTABILIZATION

Plants reduce pollutant leaching and land erosion as well as providing additional economic values, such as wood, bioenergy, dust control, and ecological services (Robinson *et al.*, 2009). On a mine tailing site in northern Chile, approximately 70 colonizing plant species had potential economic values for various uses (Orchard *et al.*, 2009). In addition, in moderately metal-contaminated agricultural soils, profitable crops have been cultivated successfully for bioenergy production (Fässler *et al.*, 2011; Greger and Landberg, 1999).

It is ideal to select and use native plant species for phytostabilization, rather than introducing potentially invasive plant species to remediation sites (Mench *et al.*, 2010). Plants used in phytostabilization should have low accumulation of pollutants in shoots and high tolerance to pollutants in the soil (Butcher, 2009; Cunningham, 1995). In hostile environments, such as mine tailing deposits, plants may contribute organic material, promoting the growth of microorganisms, which in turn will transform the tailings into a more soil-like structure (Mendez and Maier, 2008b). Mycorrhizae may also help in maintaining viable plants by increasing the plants' phosphorous uptake and reducing the As accumulation through the efflux of arsenite from the mycorrhizae into the soil solution (Sharples *et al.*, 2000). Phytostabilization will be less successful if rainfall exceeds evapotranspiration, a problem that can be addressed by recirculating the drainage water (Robinson *et al.*, 2009). On the other hand, if rainfall is too scarce, for example, in semi-arid/arid regions, supplementary irrigation may be needed at the onset of phytostabilization (Mendez and Maier, 2008a).

Trees are often promoted in phytostabilization due to their high rates of evapotranspiration, which reduce the flow of water through the soil, leading to a reduced leaching of pollutants to the surrounding surface and groundwaters (Pulford and Watson, 2003). Due to the specific characteristics of certain tree species, such as oak (*Quercus* sp.), which increases soil acidification, and poplar (*Populus* sp.), which accumulates high levels of certain metals in its leaves, these species should not be selected for phytostabilization in areas where such characteristics may result in increased mobility of pollutants (Mertens *et al.*, 2007). If the contaminants are situated below the root depth, plants will be unable to exert their phytostabilizing effect. To reach the contaminants, plants, such as trees, able to send roots to greater depths are required. However, deep rooting at desirable depths may not always develop, for example, if physical obstacles restrict root growth or if surface soil water is abundant, making the plant reluctant to spend energy trying to send its roots deeper (Negri *et al.*, 2003). In addition, if surface soil contains enough nutrients, deep root growth does not occur (Stoltz and Greger, 2006).

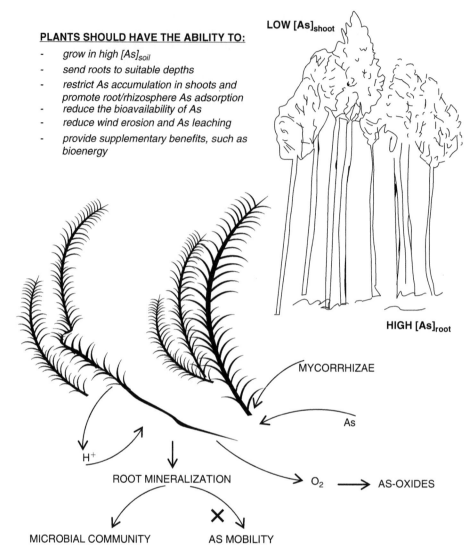

PLANTS SHOULD HAVE THE ABILITY TO:

-   *grow in high [As]$_{soil}$*
-   *send roots to suitable depths*
-   *restrict As accumulation in shoots and*
    promote root/rhizosphere As adsorption
-   *reduce the bioavailability of As*
-   *reduce wind erosion and As leaching*
-   *provide supplementary benefits, such as*
    bioenergy

LOW [As]$_{shoot}$

HIGH [As]$_{root}$

MYCORRHIZAE

As

H$^+$

ROOT MINERALIZATION

O$_2$ ⟶ AS-OXIDES

MICROBIAL COMMUNITY

AS MOBILITY

Figure 3.4.   Phytostabilization of As includes low [As]$_{shoot}$, the accumulation and adsorption of As to the roots, and immobilization through the formation of As oxides (mainly Fe, Mn, and Al) in the root zone. The pH should be kept around neutral to reduce the availability of As. Root mineralization should enhance the microorganism community but not increase As mobility. Mycorrhizae increase the plant phosphorous uptake and help regulate the symplastic As accumulation through As efflux.

3.5   PHYTOSTABILIZATION OF ARSENIC

The general mechanisms and required plant characteristics for the phytostabilization of As-polluted soils, as summarized in Figure 3.4, include an ability to grow and reach suitable root depths in soils contaminated with high concentrations of As. Furthermore, the figure shows that suitable plants should reduce As leaching/availability and wind erosion. In addition, supplementary benefits, preferably economic, as is the case with bioenergy crops, should be favored (Robinson *et al.*, 2009).

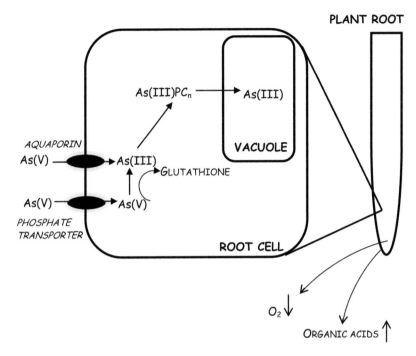

Figure 3.5.   Possible mechanisms for mobilizing or immobilizing As. The mobility of As is influenced by plants, by either the release of O_2 (decreased mobility $\downarrow$) and organic acids (increased mobility $\uparrow$) or uptake by roots. Arsenate [As(V)] is taken up and reduced to arsenite [As(III)] by glutathione. Arsenite is bound to phytochelatins (PC) and transported to the vacuole (modified from Tripathi *et al.*, 2007).

3.5.1   *Immobilization and mobilization of arsenic by plants*

To achieve successful phytostabilization, the selected plants need to express the proper character-istics in the soil–root interface to immobilize As in the contaminated environment. Plant species capable of apoplastic immobilization of As, intracellular immobilization of As by symplastic transport into root cells, and immobilizing As in the rhizosphere will be good candidates for As phytostabilization. The mechanisms for As immobilization and mobilization by plant roots are summarized in Figure 3.5.

Immobilization of As includes As accumulation in plant tissues via cellular uptake of As and storage in the vacuole. Symplastic uptake of As differs between As species. Arsenate is taken up by high-affinity phosphate transporters, while arsenite is taken up through aquaporins also used to transport small neutral molecules, such as glycerol and silicic acid (Meharg and Macnair, 1992; Zhao *et al.*, 2010). In addition, monomethylarsonic acid and dimethylarsinic acid are taken up by plants through aquaporins (Azizur Rahman *et al.*, 2011a). Inside the plant cells, various forms of As are reduced to arsenite, bound to phytochelatins, and stored in the vacuole (Tripathi *et al.*, 2007).

The immobilization of As by plants also comprises absorption and adsorption processes in the rhizosphere. For example, As is absorbed through the root apoplast or it can be adsorbed to soil particles and plant roots. Decreased mobility of As in soil using plants was demonstrated with *Lupinus albus* (Vázquez *et al.*, 2006). The apoplastic storage of As in plant roots can make up a large fraction of the total As in plants. Approximately 60% of the total As in rice (*Oryza sativa*) grown under reducing conditions was found in the root apoplast (Bravin *et al.*, 2008). Iron plaque that forms on the roots of wetland plants due to oxidation in the rhizosphere may also serve as an illustration of this. For example, the mobility of As was reduced due to As adsorption to iron plaque formed in the oxidized rhizosphere of salt marsh plants (Doyle and Otte, 1997). In addition, rice cultivars with high oxygen release ability from their roots accumulated less As in the straw and grain than did cultivars with low oxygen release ability, probably due to the formation of iron plaque with concomitant As adsorption to the plaque (Mei *et al.*, 2009).

Plants have active mechanisms for pumping organic acids into the rhizosphere to render phosphorus available. Due to chemical similarities to phosphate, arsenate may also become more available due to these mechanisms, for example, plant strategies to attack iron oxides/hydroxides (Moreno-Jiménez *et al.*, 2012). High levels of organic soil acids, such as phytic acid ($380 \,\mu g \, g^{-1}$ root dry weight) and oxalic acid ($50 \,\mu g \, g^{-1}$ root dry weight), released from plants increase the mobility of As in soils contaminated with chromated copper arsenate and/or aluminum and iron arsenates (Tu *et al.*, 2004). Due to its anionic nature, competition between As and organic molecules containing phenol, -OH, and -COOH for sorption sites on soil particles leads to increased As mobility (Grafe *et al.*, 2001). Hence, plants releasing high levels of organic acids in root exudates are unsuitable for As phytostabilization.

3.5.2   *Plant species suitable for arsenic phytostabilization*

The general characteristics of plants suitable for As phytostabilization include a high As tolerance, a low translocation factor (i.e., $[As]_{shoot}:[As]_{root}$), and a self-sustaining existence. Several plant species have been reported as interesting for As phytostabilization, and many of them were found growing naturally in mine tailings. These include *Eucalyptus*, especially *E. cladocalyx*, where no effect on As availability or soil pH was observed in tailings treated with this plant, topsoil, and biosolid amendments compared with untreated tailings (King *et al.*, 2008). Three *Viola* species growing at an abandoned mine site in Macedonia were found suitable for phytostabilization, exhibiting a low As translocation factor (<0.2) (Stefanovic *et al.*, 2010). At an old gold mine in New Zealand, several plant species, naturally occurring in the area, were found suitable for phytostabilization purposes, including local shrubs but also grasses such as *Agrostis capillaries* and *Anthoxanthum odoratum* and the reed *Juncus articulatus* (Craw *et al.*, 2007). *Rhododendron tomentosum* and *Veronica beccabunga* occurring on mine tailings in Sweden had high root accumulation ($[As]_{root}:[As]_{soil} > 2.5$) and low root-to-shoot translocation (<0.2) (Bergqvist and Greger, 2012). In an abandoned tungsten mining area in Spain, *Salix atrocinerea* was found growing in the most polluted areas with low accumulation of As in the shoots ($<7 \, mg \, As \, kg^{-1}$), while *Scirpus holoschoenus* accumulated more than $3000 \, mg \, kg^{-1}$ of As in the roots (Otones *et al.*, 2011).

Other examples come from plants growing in soil polluted from the accident at the Aznalcóllar pyrite mine in Southern Spain. *Lupinus albus* contained $3.75 \, mg \, As \, kg^{-1}$ in the shoots and $40 \, mg \, As \, kg^{-1}$ in the roots while reducing the soluble As fractions in the soil compared with soil without *Lupinus albus* (Vázquez *et al.*, 2006). Reduced availability of As in the soil was also demonstrated for *Retama sphaerocarpa*, which had a high survival rate and low shoot As accumulation ($1.5 \, mg \, As \, kg^{-1}$) (Moreno-Jiménez *et al.*, 2011).

Whole plant families thought likely to be successful in As phytostabilization include Asteraceae and Chenopodiaceae (Mendez and Maier, 2008b). Table 3.1 summarizes the plant species and families suggested as suitable for As phytostabilization.

3.6   AMENDMENTS FOR ENHANCED ARSENIC STABILIZATION

3.6.1   *Amendments for arsenic stabilization*

The addition of amendment materials can help promote conditions for successful As phytostabilization in soil. Iron-based amendments reduce As mobility (Kumpiene *et al.*, 2008). For example, in brownfield soil, the availability of As was reduced by 98% using a water treatment residue consisting mainly of ferrihydrite; furthermore, aging the residue for 103 days did not affect

Table 3.1. Plant species and families with a low [As]$_{shoot}$ and a high [As]$_{root}$, suggesting that they are suitable for As phytostabilization.

Plant family	Plant species	Reference
Asteraceae	*Baccharis neglecta*	Mendez and Maier (2008b)
Cyperaceae	*Scirpus holoschoenus*	Otones *et al.* (2011)
Chenopodiaceae	*Atriplex lentiformis*	Mendez and Maier (2008b)
Ericaceae	*Rhododendron tomentosum*	Bergqvist and Greger (2012)
Fabaceae	*Lupinus albus*	Vázquez *et al.* (2006)
Fabaceae	*Retama sphaerocarpa*	Moreno-Jiménez *et al.* (2011)
Fabaceae	*Ulex europaeus*	Craw *et al.* (2007)
Griseliniaceae	*Griselinia littoralis*	"
Myrtaceae	*Eucalyptus cladocalyx*	King *et al.* (2008)
Myrtaceae	*Leptospermum scoparium*	Craw *et al.* (2007)
Onagraceae	*Fuchsia excorticate*	"
Poaceae	*Agrostis capillaries*	"
Poaceae	*Anthoxanthum odoratum*	"
Scrophulariaceae	*Veronica beccabunga*	Bergqvist and Greger (2012)
Salicaceae	*Salix atrocinerea*	Otones *et al.* (2011)
Violaceae	*Viola allchariensis*	Stefanovic *et al.* (2010)
"	*Viola arsenica*	"
"	*Viola macedonia*	"

Table 3.2. Concentration of As [mg L^{-1}] in leachates before and after treatment with organic matter amendments (±SE).

Soil	Organic matter amendment	As (DW) [mg kg^{-1}]		Reference
		Before treatment	After treatment	
Mine tailings	Compost	0.089 ± 0.06	0.424 ± 0.05	Renella *et al.* (2008)
"	Sewage sludge, unlimed	18.78 ± 3.62	17.68 ± 2.20	Stolz and Greger (2002)
"	Sewage sludge, limed	19.14 ± 2.76	64.10 ± 12.01	Stolz and Greger (2002)
Brownfield	Compost	~8	~55	Clemente *et al.* (2010)
Agriculture	Compost	1.84	1.91	De La Fuente *et al.* (2010)

these results (Nielsen *et al.*, 2011). Similarly, amendments rich in iron oxides, based on by-products from the steel industry, reduced the availability of As in low-As-polluted agricultural soils (0–20 mg As kg^{-1}) (De La Fuente *et al.*, 2010; Gutierrez *et al.*, 2010). These amendments are all relatively cheap, especially the steel industry by-products. If large areas, for example, of agricultural land, need to be amended, the cost-effectiveness of the project is important. The cost and efficiency of an amendment will determine whether its application is feasible. Calcium-based amendments have also been found to significantly reduce the amount of plant-available As (Gutierrez *et al.*, 2010).

3.6.2   *Unsuitable or inefficient amendments for arsenic stabilization*

As opposed to iron-based amendments, additions of organic matter generally increase or do not affect the mobility of As. For example, the addition of biosolids increased the leaching of As from neutral mine tailings (Pond *et al.*, 2005). Phytostabilization of a former mine site using compost, beringite, and zerovalent iron grit amendments proved successful for Cd, Pb, Mn, and Zn, but not for As, the mobility of which was higher in the treated than the untreated site (Renella *et al.*, 2008) (Table 3.2). The authors conclude that a possible explanation for the increase in As mobility could be the desorption of As from iron oxides due to competition from organic matter. Similarly, amendments consisting of sewage sludge and fly ash from wood combustion increased the mobility of As in drainage water from limed mine tailings (Stolz and Greger, 2002). In soil from an old alkali works contaminated with As, Cu, and Pb, amendments consisting of green waste compost led to an increase in As mobility with a concomitant increase in uptake of As in lettuce and sunflower (Clemente *et al.*, 2010). In an agricultural soil, amendments of organic matter did not reduce the mobility of As (De La Fuente *et al.*, 2010). Table 3.2 summarizes the concentrations of As in leachates from soil before and after treatments with organic matter amendments.

Treating As-polluted soil with phosphate and oxalic acid leads to a drastic increase in the mobility of As. Phosphate and oxalic acid mobilize As by inhibiting the adsorption of arsenate and arsenite and by dissolving As-associated aluminum and iron oxides/hydroxides in the soil, respectively (Wovkulich *et al.*, 2010). In agricultural soil, adding an inorganic nutrient fertilizer including nitrogen, phosphorous, and potassium (15:15:15 N:P:K) increased As mobility (De La Fuente *et al.*, 2010). In soil contaminated with As and Pb, phosphate amendments, such as calcium-magnesium-phosphate and rock phosphate, reduced the Pb mobility but increased the As mobility, while ferrous sulfate amendments produced the opposite results (Cui *et al.*, 2010). Phosphate is a widely accepted amendment for immobilizing Pb in soil and aqueous solutions (Miretzky and Fernandez Cirelli, 2009). These results highlight the problems of treating sites with multiple contaminants and the need to establish individual strategic management plans for each individual site.

3.7   MANAGEMENT PLAN FOR ARSENIC PHYTOSTABILIZATION

How should phytostabilization in an As-polluted area be established to successfully encourage both plant vegetation and As immobility? Soil quality and pH may have to be adjusted using amendments to promote the hospitability and the desired physicochemical properties of the soil. Plant species with desired abilities should be selected. If phytostabilization strategies are unable to reduce the mobility of As and/or other contaminants, additional methods along with phytostabilization may be employed to obtain effective stabilization of the contaminants. In Section 3.7, we will discuss suggestions for a management plan for successful As phytostabilization.

3.7.1   *Soil parameters that influence arsenic mobility*

The first step when establishing As phytostabilization in an area is to characterize the soil to determine the risks of As mobility and what actions to take. For example, adsorption of As differs between soils, for example, As adsorption is higher on clayey than sandy soil (Kumpiene *et al.*, 2008). Characteristics such as high soil organic matter ($>25\%$) may lead to increased As mobility, possibly due to the desorption of As by dissolved organic matter (Mench *et al.*, 2003). Amendments could be added to address problems of high As mobility due to the physicochemical properties of the soil (see Section 3.7.2 below).

To prevent increased As mobility due to anionic competition between hydroxyl ions and As, and protonation of arsenate (pH < 2.5), the soil pH should be neither alkaline nor acidic (Moreno-Jiménez *et al.*, 2012). A specific pH interval for optimal stabilization is impossible to determine due to the specific characteristics of each soil; however, As sorption on solid phases such as Fe oxides is promoted in the pH range around neutral (Renella *et al.*, 2008). The pH could be adjusted by adding amendments such as carbonates or humic substances (Moreno-Jiménez *et al.*, 2012).

Redox potential should be high (p$E > 10$) to ensure that the dominant As species is the more strongly adsorbed arsenate and not the more mobile arsenite (Zhao *et al.*, 2010). Redox potential could be increased by diverting and reducing the water flow and the groundwater level in the area.

Table 3.3.   Soil and plant species parameters for successful As phytostabilization.

Soil/plant parameters	Setting	Reference
Amendments	Fe-based, MnO$_x$, AlO$_x$ alkaline, clay	Kumpiene *et al.* (2008)
Organic matter	Below 10% (possibly associated with a sealing layer)	Mench *et al.* (2003); Neuschütz and Greger (2010)
pH	7	Renella *et al.* (2008)
Phosphate	<1 mM	Wovkulich *et al.* (2010)
Plant species	Root mineralization supports As immobilization	Moreno-Jiménez *et al.* (2009)
Redox potential	pE > 10	Zhao *et al.* (2010)

3.7.2   *Amendments that encourage plant vegetation and As immobility*

Amendments, such as iron compounds, manganese and aluminum oxides, alkaline materials, and clay minerals, could be added to reduce the mobility of As in soils (Kumpiene *et al.*, 2008) (see Section 3.6.1). By-products from industrial manufacturing can be ideal materials for As soil remediation. For example, iron-rich industrial by-products have proven to successfully reduce the availability of As and metals from a contaminated gold mine area (Lee *et al.*, 2011).

In soils that are poor in quality in terms of nutrients and phytotoxic levels of trace metals and other contaminants, such as mine wastes, organic matter may have to be added to allow plant growth for phytostabilization (Moreno-Jiménez *et al.*, 2010). Amendments added to soil have to be adjusted to maintain low As availability and to improve soil quality and stability. Organic matter amendments may increase As mobility (see Section 3.6.2), but a combination of organic matter and iron-based amendments could provide a good alternative for achieving this purpose (Moreno-Jiménez *et al.*, 2010). An alternative for reducing the addition of organic matter, while still promoting a habitable soil for plants, is the vertical separation of the soil into different strata. The addition of a sealing layer between the As-contaminated soil and a vegetative cover could reduce the need to add organic matter to the soil. For example, pyrite mine tailings were covered with a sealing layer composed of fly ash from wood combustion, followed by a layer of sewage sludge; this vertical separation reduced the mobility of both nutrients and metals (Neuschütz and Greger, 2010).

3.7.3   *Selecting plant species for arsenic phytostabilization*

The selection of plant species for As phytostabilization should ensure that the plant species does not increase the As mobility from the area and that it is suitable for As phytostabilization in terms of As accumulation (see Section 3.5.2). For example, Moreno-Jiménez *et al.* (2009) demonstrated that root mineralization in five species of shrubs, i.e., *Arbutus unedo*, *Myrtus comunnis*, *Pistacia lentiscus*, *Retama sphaerocarpa*, and *Rosmarinus officinalis*, did not increase As mobility, supporting the idea of successful As phytostabilization. However, one plant species, *Tamarix gallica*, promoted higher release of As from the soil after mineralization, compared with the control soil, highlighting the importance of plant-specific characteristics and of selecting proper plant species for phytostabilization (Moreno-Jiménez *et al.*, 2009). Table 3.3 summarizes some of the soil and plant parameters that can be used when developing management plans for successful As phytostabilization.

3.7.4   *Methods suitable for combining with arsenic phytostabilization*

Phytofiltration could be employed as an additional strategy in combination with phytostabilization in situations in which phytostabilization is insufficient to reduce As mobility. This could occur in areas with multiple contaminants, a situation common in polluted areas. Treatments that reduce heavy metal mobility, for example, organic matter amendments, might at the same time increase As mobility (Clemente *et al.*, 2010; Pond *et al.*, 2005; Stolz and Greger, 2002). The costs of combining amendments or of achieving optimal physicochemical conditions to phytostabilize all pollutants in an area might be un-realistic. Phytostabilization in combination with phytofiltration could prove a cost-effective and realistic alternative. Several plant species, for example, duckweed (*Lemna* sp., *Spirodela* sp., and *Wolffia* sp.), watercress (*Nasturtium officinale* and *Nasturtium microphyllum*), water hyacinth (*Eichornia crassipes*), water fern (*Azolla* sp.), and Hydrilla (*Hydrilla verticillata*), have produced promising results in the phytofiltration of As (Azizur Rahman and Hasegawa, 2011b). In addition, Canadian waterweed (*Elodea Canadensis*) could remove up to 75% of the As from polluted water (Greger *et al.*, 2010).

3.8   CONCLUDING REMARKS

Phytostabilization of As has not always produced satisfactory results when strategies for phytostabilizing heavy metals have been employed, such as adding organic matter amendments and introducing plants (Renella *et al.*, 2008; Stolz and Greger, 2002). Instead, management plans that combine As phytostabilization parameters could be used as tool to obtain successful results. For example, specific attention must be paid to As soil chemistry, including maintenance of a high redox potential ($pE > 10$) and a pH of approximately 7. The use of organic matter to improve poor-quality soil should be used in combination with other amendments, for example, iron-based ones, to reduce the risk of increased As mobility. Specific consideration should be given to the recycling of nutrients in the phytostabilized area to promote self-sustaining, long-term pollutant control. Adding nutrients to the phytostabilized area, in the form of either compost or inorganic nutrients, may lead to increased As mobility due to anionic competition (Clemente *et al.*, 2010; De La Fuente *et al.*, 2010). If multiple contaminants need to be dealt with, which is often the case, plants successfully tested for As phytostabilization might fail to control As in the long term by reducing its mobility. In sites with multiple contaminants, capping along with the planting of shallow-rooted plants may be preferable to prevent leaching of both metals and As. Another option is to establish deep-rooted plants that do not increase As mobility upon mineralization (Moreno-Jiménez *et al.*, 2009).

If the above parameters for successful As phytostabilization cannot be met, and As remains mobile, phytofiltration could be used to supplement phytostabilization. Organic matter amendments can be used for the phytostabilization of heavy metals. Thereafter, the problem of increased As mobility due to the additional organic matter can be addressed by employing phytofiltration.

REFERENCES

Ackermann, J., Vetterlein, D., Kaiser, K., Mattusch, J. & Jahn, R.: The bioavailability of arsenic in floodplain soils: a simulation of water saturation. *Eur. J. Soil Sci.* 61 (2010), pp. 84–96.

Andreae, M.O. & Klumpp, D.: Biosynthesis and release of organoarsenic compounds by marine algae. *Environ. Sci. Technol.* 13:6 (1979), pp. 738–741.

Azizur Rahman, M. & Hasegawa, H.: Aquatic arsenic: Phytoremediation using floating macrophytes. *Chemosphere* 83 (2011b), pp. 633–646.

Azizur Rahman, M., Kadohashi, Maki, T.K. & Hasegawa, H.: Transport of DMAA and MMAA into rice (*Oryza sativa* L.) roots. *Environ. Exp. Bot.* 72 (2011a), pp. 41–46.

Bergqvist, C. & Greger, M.: Arsenic accumulation and speciation in plants from different habitats. *Appl. Geochem.* 27 (2012), pp. 615–622.

Bhumbla, D.K. & Keefer, R.F.: Arsenic mobilization and bioavailability in soils. In: J.O. Nriagu (ed): *Arsenic in the environment*, Part I: *Cycling and Characterization*. Wiley, New York, 1994, pp. 51–58.

Bowell, R.J.: Sulphide oxidation and arsenic speciation in tropical soils. *Environ. Geochem. Health* 16 (1994), pp. 84–91.

Bravin, M.N., Travassac, F., Le Floch, M., Hinsinger, P. & Garnier, J.-M.: Oxygen input controls the spatial and temporal dynamics of arsenic at the surface of a flooded paddy soil and in the rhizosphere of lowland rice (*Oryza sativa* L.): a microcosm study. *Plant Soil* 312 (2008), pp. 207–218.

Brown, S., Sprenger, M., Maxemchuk, A. & Compton, H.: Ecosystem function in alluvial tailings after biosolids and lime addition. *J. Environ. Qual.* 34 (2005), pp. 139–148.

Butcher, D.J.: Phytoremediation of arsenic: fundamental studies, practical applications, and future prospects. *Appl. Spectrosc. Rev.* 44 (2009), pp. 534–551.

Cao, X., Ma, L.Q. & Shiralipour, A.: Effects of compost and phosphate amendments on arsenic mobility in soils and arsenic uptake by the hyperaccumulator, *Pteris vittata* L. *Environ. Pollut.* 126 (2003), pp. 157–167.

Clemente, R., Hartley, W., Riby, P., Dickinson, N.M. & Lepp, N.W.: Trace element mobility in a contaminated soil two years after field-amendment with a greenwaste compost mulch. *Environ. Pollut.* 158 (2010), pp. 1644–1651.

Commission Directive 2003/40/EC. *Official Journal of the European Union*, L 126/34, EN, 22.5.2003.

Craw, D., Rufaut, C., Haffert, L. & Paterson, L.: Plant colonization and arsenic uptake on high arsenic mine wastes, New Zealand. *Water Air Soil Poll.* 179 (2007), pp. 351–364.

Cui, Y., Du, X., Weng, L. & Van Riemsdijk, W.H.: Assessment of in situ immobilization of lead (Pb) and arsenic (As) in contaminated soils with phosphate and iron: solubility and bioaccessibility. *Water Air Soil Poll.* 213 (2010), pp. 95–104.

Cunningham, S.D., Berti, W.R. & Huang, J.W.: Phytoremediation of contaminated soils. *Trends Biotechnol.* 13:9 (1995), pp. 393–397.

De La Fuente, C., Clemente, R., Alburquerque, J.A., Vélez, D. & Bernal, M.P.: Implications of the use of As-rich groundwater for agricultural purposes and the effects of soil amendments on As solubility. *Environ. Sci. Technol.* 44 (2010), pp. 9463–9469.

Dopp, E., von Recklinghausen, U., Diaz-Bone, R., Hirner, A.V. & Rettenmeier, A.W.: Cellular uptake, subcellular distribution and toxicity of arsenic compounds in methylating and non-methylating cells. *Environ. Res.* 110 (2019), pp. 435–442.

Doyle, M.O. & Otte, M.L.: Organism-induced accumulation of iron, zinc and arsenic in wetland soils. *Environ. Pollut.* 96:1 (1997), pp. 1–11.

EFSA: Scientific opinion on arsenic in food. *EFSA Journal* 7:10 (2009), p. 1351.

Fässler, E., Robinson, B.H., Stauffer, W. Gupta, S.K., Papritz, A. & Schulin, R.: Phytomanagement of metal-contaminated agricultural land using sunflower, maize and tobacco. *Agr. Ecosyst. Environ.* 136 (2010), pp. 49–58.

Grafe, M., Eick, M.J. & Grossl, P.R.: Adsorption of arsenate(V) and arsenite(III) on goethite in the presence and absence of dissolved organic carbon. *Soil Sci. Soc. Am. J.* 65 (2001), pp. 1680–1687.

Greger, M. & Landberg, T.: Use of willow in phytoextraction. *Int. J. Phytorem.* 1: (1999), pp. 115–123.

Greger, M., Sandhi, A., Nordstrand, D., Bergqvist, C. & Nyquist-Rennerfelt, J.: Water leaking from toxic elements using phytofiltration with *Elodea Canadensis*. *Proceeding, METEAU, COST Action* 637, Kristianstad, Sweden, 2010.

Gutierrez, J., Hong, C.O., Lee, B-H. & Kim, P.J.: Effect of steel-making slag as a soil amendment on arsenic uptake by radish (*Raphanus sativa* L.) in an upland soil. *Biol. Fert. Soils* 46 (2010), pp. 617–623.

Hartley, W., Dickinson, N.M., Clemente, R., French, C. Piearce, T.G., Sparke, S. & Lepp, N.W.: Arsenic stability and mobilization in soil at an amenity grassland overlying chemical waste (St. Helens, UK). *Environ. Pollut.* 157 (2009), pp. 847–856.

Kaize, T., Watanbe, S. & Itoh, K.: The acute toxicity of arsenobetaine. *Chemosphere* 14:9 (1985), pp. 1327–1332.

King, D.J., Doronila, A.I., Feenstra, C., Baker, A.J.M. & Woodrow, I.E.: Phytostabilisation of arsenical gold mine tailings using four *Eucalyptus* species: Growth, arsenic uptake and availability after five years. *Sci. Total Environ.* 406 (2008), pp. 35–42.

Kumpiene, J., Lagerkvist, A. & Maurice, C.: Stabilization of As, Cr, Cu, Pb and Zn in soil using amendments – A review. *Waste Manag.* 28 (2008), pp. 215–225.

Lee, S.-H., Kim, E.Y., Park, H., Yun, J. & Kim, J.-G.: In situ stabilization of arsenic and metal-contaminated agricultural soil using industrial by-products. *Geoderma* 161 (2011), pp. 1–7.

Ma, Y., Lombi, E., Oliver, I.W., Nolan, A.L. & McLaughlin, M.J.: Long-term aging of copper added to soils. *Environ. Sci. Technol.* 40 (2006), pp. 6310–6317.

McBride, B.C. & Wolfe, R.S.: Biosynthesis of dimethylarsine by methanobacterium. *Biochemistry-US* 10:23 (1971), pp. 4312–4317.

Meharg, A.A. & Hartley-Whitaker, J.: Arsenic uptake and metabolism in arsenic resistant and nonresistant plant species. *New Phytol.* 154 (2002), pp. 29–43.

Meharg, A.A. & Jardine, L.: Arsenite transport into paddy rice (*Oryza sativa*) roots. *New Phytol.* 157 (2003), pp. 39–44.

Meharg, A.A. & Macnair, M.R.: Suppression of the high affinity phosphate uptake system: A mechanism of arsenate tolerance in *Holcus lanatus* L. *J. Exp. Bot.* 43:249 (1992), pp. 519–524.

Mei, X.Q., Ye, Z.H. & Wong M.H.: The relationship of root porosity and radial oxygen loss on arsenic tolerance and uptake in rice grains and straw. *Environ. Pollut.* 157 (2009), pp. 2550–2557.

Mench, M., Vangronsveld, J., Clijsters, H., Lepp, N.W. & Edwards, R.: In situ metal immobilization and phytostabilization of contaminated soils. In: N. Terry & G. Banuelos (eds): *Phytoremediation of contaminated soil and water*. CRC Press LLC, Boca Raton, FL, 2000.

Mench, M., Bussière, S., Boisson, J., Castaing, E., Vangronsveld, J., Ruttens, A., De Koe, T., Bleeker, P., Assuncão, A. & Manceau, A.: Progress in remediation and revegetation of the barren Jales gold mine spoil after in situ treatments. *Plant Soil* 249 (2003), pp. 187–202.

Mench, M., Lepp, N., Ber, V., Schwitzgübel, J.-P., Gawronski, S., Schröder, W.: Successes and limitations of phytotechnologies at field scale: outcomes, assessment and outlook from COST Action 859. *J. Soils Sediments* 10 (2010), pp. 1039–1070.

Mendez, M.O. & Maier, R.M.: Phytostabilization of mine tailings in arid and semiarid environments – an emerging remediation technology. *Environ. Health Persp.* 116 (2008a), pp. 278–283.

Mendez, M.O. & Maier, R.M.: Phytoremediation of mine tailings in temperate and arid environments. *Rev. Environ. Sci. Biotechnol.* 7 (2008b), pp. 47–59.

Mertens, J., Van Nevel, L., De Schrijver, A., Piesschaert, F., Oosterbaan, A., Tack, F.M.G. & Verheyen, K.: Tree species effect on the redistribution of soil metals. *Environ. Pollut.* 149 (2007), pp. 173–181.

Mestrot, A., Feldmann, J., Krupp, E.M., Hossain, M.S., Roman-Ross, G. & Meharg, A.A.: Field fluxes and speciation of arsines emanating from soils. *Environ. Sci. Technol.* 45 (2011), pp. 1798–1804.

Meunier, L., Koch, I. & Reimer K.J.: Effects of organic matter and ageing on the bioaccessibility of arsenic. *Environ. Pollut.* 159 (2011), pp. 2530–2536.

Mikutta, C. & Kretzschmar, R.: Spectroscopic evidence for ternary complex formation between arsenate and ferric iron complexes of humic substances. *Environ. Sci. Technol.* 45 (2011), pp. 9550–9557.

Miretzky, P. & Fernandez Cirelli, A.: Phosphates for Pb immobilization in soils: a review. Org. Farm Pest. *Contr. Rem. Soil Poll.* 1 (2009), pp. 351–370.

Moreno-Jiménez, E., Peñalosa, J.M., Esteban, E. & Bernal, M.P.: Feasibility of arsenic phytostabilisation using Mediterranean shrubs: impact of root mineralization on As availability in soils. *J. Environ. Monitor.* 11 (2009), pp. 1375–1380.

Moreno-Jiménez, E., Manzano, R., Esteban, E. & Peñalosa, J.: The fate of arsenic in soils adjacent to an old mine site (Bustarviejo, Spain): mobility and transfer to native flora. *J. Soils Sediments* 10 (2010), pp. 301–312.

Moreno-Jiménez, E., Vázquez, S., Carpena-Ruiz, R.O., Esteban, E. & Peñalosa, J.M.: Using Mediterranean shrubs for the phytoremediation of a soil impacted by pyritic wastes in Southern Spain: A field experiment. *J. Environ. Manage.* 92 (2011), pp. 1584–1590.

Moreno-Jiménez, E., Esteban, E. & Peñalosa, J.M.: The fate of arsenic in soil-plant systems. *Rev. Environ. Contam. Toxicol.* 215 (2012), pp. 1–37.

Negri, M.C., Gatliff, E.G., Quinn, J.J. & Hinchman, R.R.: Root development and rooting at depths. In: S.C. McCutcheon & J.L. Schnoor (eds): *Phytoremediation transformation and control of contaminants*. John Wiley & Sons, Inc., New Jersey, 2003.

Neuschütz, C. & Greger, M.: Stabilization of mine tailings using fly ash and sewage sludge planted with *Phalaris arundinacea* L. *Water Air Soil Poll.* 207 (2010), pp. 357–367.

Nielsen, S.S., Petersen, L.R., Kjeldsen, P. & Jakobsen, R.: Amendment of arsenic and chromium polluted soil from wood preservation by iron residues from water treatment. *Chemosphere* 84 (2011), pp. 383–389.

Orchard, C., León-Lobos, P. & Ginocchio, R.: Phytostabilization of massive mine wastes with native phyto-genetic resources: potential for sustainable use and conservation of the native flora in north-central Chile. *Ciencia e Investigacion Agraria* 36 (2009), pp. 329–352.

Otones, V., Álvarez-Ayuso, E., García-Sánchez, A., Santa Regina, I. & Murciego, A.: Arsenic distribution in soils and plants of an arsenic impacted former mining area. *Environ. Pollut.* 159 (2011), pp. 2637–2647.

Pond, A.P., White, S.A, Milczarek, M. & Thompson, T.L.: Accelerated weathering of biosolid-amended copper mine tailings. *J. Environ. Qual.* 34 (2005), pp. 1293–1301.

Pulford, I.D. & Watson, C.: Phytoremediation of heavy metal-contaminated land by trees – a review. *Environ. Int.* 29 (2003), pp. 529–540.

Raven, K.P., Jain, A. & Loeppert, R.H.: Arsenite and arsenate adsorption on ferrihydrite: kinetics, equilibrium, and adsorption envelopes. *Environ. Sci. Technol.* 32 (1998), pp. 344–349.

Renella, G., Landi, L., Ascher, J., Ceccherini, M.T., Pietramellara, G., Mench, M. & Nannipieri, P.: Long-term effects of aided phytostabilization of trace elements on microbial biomass and activity, enzyme activities, and composition of microbial community in the Jales contaminated mine spoils. *Environ. Pollut.* 152 (2008), pp. 702–712.

Robinson, B.H., Green, S.R., Chancerel, B., Mills, T.M. & Clothier, B.: Poplar for the phytomanagement of boron contaminated sites. *Environ. Pollut.* 150 (2007), pp. 225–233.

Robinson, B.H., Bañuelos, G., Conesa, H.M., Evangelou, M.W.H. & Schulin, R.: The phytomanagement of trace elements in soil. *Cr. Rev. Plant Sci.* 28 (2009), pp. 240–266.

Sadiq, M.: Arsenic chemistry in soils: an overview of thermodynamic predictions and field observations. *Water Air Soil Poll.* 93 (1997), pp. 117–136.

Sharples, J.M., Meharg, A.A., Chambers, S.M. & Cairney, J.W.G.: Symbiotic solution to arsenic contamination. *Nature* 404 (2000), pp. 951–952.

Sheppard, S.C.: Summary of phytotoxic levels of soil arsenic. *Water Air Soil Poll.* 64 (1992), pp. 539–550.

Stefanovic, B., Drazic, G., Tomovic, G., Sinzar-Sekulic, J., Melovski, L., Novovic, I. & Markovic, D.M.: Accumulation of arsenic and heavy metals in some Viola species from an abandoned mine, Alchar, Republic of Macedonia (FYROM). *Plant Biosyst.* 144: 3 (2010), pp. 644–655.

Stolz, E. & Greger, M.: Cottongrass effects on trace elements in submersed mine tailings. *J. Environ. Qual.* 31 (2002), pp. 1477–1483.

Stoltz, E. & Greger, M.: Root penetration through sealing layers at mine deposit sites. *Waste Manag. Res.* 24 (2006), pp. 552–559 (errata vol. 25:392).

Tripathi, R., Srivastava, S., Mishra, S., Singh, N., Tuli, R., Gupta, D.K., & Maathuis, J.M.: Arsenic hazards: strategies for tolerance and remediation by plants. *Trends Biotechnol.* 25:4 (2007), pp. 158–165.

Tu, S., Ma, L. & Luongo, T.: Root exudates and arsenic accumulation in arsenic hyperaccumulating *Pteris vittata* and non-hyperaccumulating *Nephrolepis exaltata*. *Plant Soil* 258 (2004), pp. 9–19.

Vázquez, S., Agha, R., Granado, A., Sarro, M.J., Esteban, E., Penalosa, J.M. & Carpena, R.O.: Use of white lupin plant for stabilization of Cd and As polluted acid soil. *Water Air Soil Poll.* 177 (2006), pp. 349–365.

Wang, S. & Mulligan, C.N.: Effect of natural organic matter on arsenic mobilization from mine tailings. *J. Hazard. Mater.* 168 (2009), pp. 721–726.

Ward, O.P. & Singh, A.: Soil bioremediation and phytoremediation – An overview. In: A. Singh & O.P. Ward (eds): *Applied bioremediation and phytoremediation*. Springer-Verlag, Berlin, Germany, 2004.

Wood, J.M.: Biological cycles for toxic elements in the environment. *Science* 183 (1974), pp. 1049–1052.

Wovkulich, K., Mailloux, B.J, Lacko, A., Keimowitz, A.R., Stute, M., Simpson, H.J. & Chillrud, S.N.: Chemical treatments for mobilizing arsenic from contaminated aquifer solids to accelerate remediation. *Appl. Geochem.* 25 (2010), pp. 1500–1509.

Zhao, F.-J., McGrath, S.P. & Meharg, A.A.: Arsenic as a food chain contaminant: mechanisms of plant uptake and metabolism and mitigation strategies. *Annu. Rev. Plant Biol.* 61 (2010), pp. 535–559.

CHAPTER 4

Recent advances in phytoremediation of arsenic-contaminated soils

Xin Wang & Lena Qiying Ma

4.1 INTRODUCTION

Arsenic contamination in soils occurs widely in a range of ecosystems resulting from geological origins and anthropogenic activities. On average, arsenic concentration ranges from 5 to 10 mg kg^{-1} in uncontaminated soils and above 10 mg kg^{-1} in contaminated soils (Hossain, 2006). Increased buildup of arsenic in irrigated soils has been widely recognized in South and South-east Asia (Brammer and Ravenscroft, 2009), posing significant threats to agriculture sustainability. In Bangladesh, long-term irrigation with arsenic-rich groundwater from shallow aquifers in dry season adds >1000 tons of arsenic to the agricultural soils (Ali *et al.*, 2003). In addition, arsenic contamination in soils results from various anthropogenic activities, such as mining and smelting (Williams *et al.*, 2009), and using arsenic-containing wood preservatives (Chirenje *et al.*, 2003), pigment, pesticides, herbicide (Sarkar *et al.*, 2005) and feed additives (Arai *et al.*, 2003).

As a cost-effective and ecology-friendly technology, phytoremediation of arsenic-contaminated soils has been widely studied. Among phytoremediation technologies, phytoextraction and phytostabilization are two predominant approaches in remediation of soils contaminated with heavy metals. Phytoextraction takes advantage of plants to remove contaminants from soils by concentrating the targeted contaminant to the harvestable tissues (Salt *et al.*, 1998). To achieve effective arsenic removal from soils, the plant should be highly tolerant to arsenic and efficient in accumulating arsenic into sufficient aboveground biomass. Therefore, phytoextraction efficiency depends on both aboveground biomass yield and plant arsenic concentration. Bioconcentration factor (*BF*), which is defined as the ratio of element concentration in plant shoots to that in soil, has been used to measure a plant's efficiency in phytoextraction. Based on mass balance calculation, phytoextraction is feasible only by using plants with *BF* much greater than 1, regardless of how large the harvestable biomass (McGrath and Zhao, 2003). Furthermore, to achieve efficient removal of contaminant in a reasonable time frame with high plant survival and biomass yield, the initial and target soil contaminant concentrations should be taken into account to predict the applicability of phytoextraction, which is in most cases appropriate for soils with low contamination (Zhao and McGrath, 2009).

For heavily contaminated sites (e.g., industrial and mining degraded sites), indigenous tolerant species with extensive root system and low translocation factor (*TF*, the ratio of contaminant concentration in shoots to that in roots) provide valuable plant resources to immobilize the pollutant in the rhizosphere, and simultaneously stabilize the degraded sites by establishing vegetation cover. Soil amendments, in some cases, are essential to assist the success of the survival of pioneering species by mitigating contaminant toxicity and improving substrate conditions (Vangronsveld *et al.*, 2009). In this way, ecological restoration of contaminated sites can be gradually achieved through revegetation, which is termed as phytostabilization.

Beside these two major phytoremediation techniques, other methods include phytoexclusion and rhizofiltration. To remediate large-scale agricultural soils contaminated by arsenic, phytoexclusion is more practical to reduce arsenic transfer from soil to crops. Based on the well-established knowledge with regard to arsenic biogeochemistry and arsenic transport mechanisms in rice, a range of strategies including water management, Si fertilization, and rhizosphere manipulation have been studied to mitigate arsenic contamination in soil-rice system (Kertulis *et al.*, 2005; Ma *et al.*, 2008; Zhao *et al.*, 2010). To prevent loading of arsenic into soils from irrigating water, which has been well recognized in South and South-east Asia, arsenic phytofiltration technique, using arsenic hyperaccumulators or aquatic plants with substantial arsenic accumulating ability can be taken into account for arsenic reduction in irrigation water (Rahman *et al.*, 2007).

In this chapter, phytoextraction and phytostabilization using As hyperaccumulators and tolerant species in both greenhouse- and field-scale studies are discussed from different perspectives. With prevalent hazards of arsenic in South Asia, potential phytoexclusion strategies are proposed to mitigate arsenic contamination in a soil-rice system. Furthermore, the feasibility of using phytofiltration to reduce continuous loading of arsenic into agricultural soils is discussed.

4.2   PHYTOEXTRACTION OF ARSENIC CONTAMINATED SOILS

As the first known arsenic hyperaccumulator, *Pteris vittata* L. (Chinese brake fern) exhibits great potentials to extract arsenic efficiently from contaminated soils and concentrate >90% of the arsenic in the fronds with *BF* up to 126 and little toxicity (Ma *et al.*, 2001). Furthermore, the plant grows fast and yields considerable biomass, particularly in tropical and subtropical areas. These constitutive traits of *P. vittata* have attracted extensive attention regarding its application in phytoextraction of arsenic-contaminated sites. Most studies have demonstrated effective extraction of arsenic by *P. vittata* in a reasonable time frame with different experiment scales (Gonzaga *et al.*, 2008; Kertulis-Tartar *et al.*, 2006; Tu *et al.*, 2002). The effective decontamination of arsenic-contaminated soil by *P. vittata* is largely attributed to its unique metabolism traits, through effective mobilization of arsenic in the rhizosphere, efficient uptake by the roots and translocation to the fronds. Following the discovery of *P. vittata*, additional 11 fern species belonging to Pteridaceae family have been identified to hyperaccumulate arsenic, providing alternative plant resources for phytoextraction of arsenic from contaminated soils (Table 4.1). As detailed in this section, phytoextraction using *P. vittata* is discussed with the underlying mechanisms.

4.2.1   *Efficient arsenic extraction by P. vittata*

P. vittata is known for its effectiveness in arsenic phytoextraction from soils. In a greenhouse experiment, *P. vittata* were grown in an arsenic-contaminated soil containing 98 mg As kg^{-1}. After 20 weeks of growth, ~26% of soil arsenic from each pot (1.5 kg soil pot^{-1}) was removed

Table 4.1.   Arsenic hyperaccumulators identified up to date in Pteridaceae family.

Species	Frond arsenic concentration [mg kg^{-1} dw]	Soil arsenic [mg As kg^{-1}]	Reference
Pteris vittata	2500–22630	50–1500	Ma *et al.* (2001)
Pteris cretica	6200–7600	500	Zhao *et al.* (2002)
Pteris umbrosa			
Pteris longifolia			
Pteris biaurita	1770–3650	100	Srivastava *et al.* (2006)
Pteris quadriaurita			
Pteris ryukyuensis			
Pteris aspericaulis	4000	200	Wang *et al.* (2007)
Pteris fauriei	3275		
Pteris oshimensis	1365		
Pteris multifida	3942		
Pityrogramma calomelanos	2270–6380	20–8800	Visoottiviseth *et al.* (2002)

by *P. vittata* with frond arsenic up to 7,230 mg kg^{-1} and *BF* of 74 (Tu *et al.*, 2002). In a separate greenhouse study investigating the effect of repeated harvests on arsenic removal by *P. vittata*, ferns were grown in six arsenic-contaminated soils with total arsenic ranging from 22.7 to 640 mg kg^{-1} (Gonzaga *et al.*, 2008). With the application of extended time-release base fertilizer (N:P:K ratios of 18:6:12) at a rate of 2 g kg^{-1} soil, the ferns were well-established after 4 months of growth and produced a good frond biomass ranging from 24.8 to 33.5 g plant^{-1} with frond arsenic being 66 to 6,151 mg kg^{-1} and *BF* being 4.7 to 48 from the first harvest (October, 2003, 4 months after transplant). During the harvest, all aboveground biomass was removed, making it difficult for the plants to re-grow under a cooler climate in the second growing period. As a result, 34–75% lower frond arsenic concentrations and 40–84% less frond biomass were obtained upon the second harvest (April, 2004). The results indicate that, though *P. vittata* is effective in arsenic extraction, proper growth timing and harvest method (leaving some fiddleheads for faster regrowth) is important to achieve optimum phytoextraction.

This conclusion is supported by a two-year field study of Kertulis-Tartar *et al.* (2006) who found that it is necessary to leave the fiddleheads (young fronds) as well as few live fronds at harvest to facilitate the regeneration and survival of ferns in winter season. Furthermore, compared to senesced fronds, live fronds accumulated 25–49% higher arsenic concentrations in field conditions, suggesting the necessity of frond harvest before they senesce to maximize arsenic extraction. In the field trial in North Central Florida, surface soil arsenic was reduced from 190 to 140 mg kg^{-1} by *P. vittata* after 2 years with planting density of 0.09 m^2 per fern (Kertulis-Tartar *et al.*, 2006). Based on the projected removal capacity, 7–8 years is needed to remove arsenic in top 15 cm soil below the cleanup level for residential site (2.1 mg kg^{-1}) or commercial site (12 mg kg^{-1}) established by Florida Department of Environmental Protection. If reasonable remediation time can be achieved using *P. vittata*, in some cases, phytoextraction could be competitive in comparison with other conventional technologies for soil remediation (Table 4.2).

The applicability of phytoextraction using *P. vittata* has been further tested in 21 contaminated-soils from England with different soil types, arsenic contamination sources and concentrations as well as the coexistence of Cu, Cd, Zn and Pb (Shelmerdine *et al.*, 2009). After three sequential growth and harvest of *P. vittata* over 9 months, 0.1–13% of total soil arsenic was removed from the soils where arsenic removal efficiency varied by up to 130-fold. Higher arsenic depletion was found under conditions with relatively low available P and low contamination of Pb, Cd, Cu and Zn. For instance, with comparable total arsenic being in soil-1 (367 mg kg^{-1}) and soil-4 (330 mg kg^{-1}), 1.7% of soil arsenic was removed after three harvests of fern fronds from soil-1 compared to only 0.26% from soil-4, which contained 3–24 times higher concentrations of Pb, Cd, Cu and Zn than soil-1. This confirmed that *P. vittata* performed better in soils with only arsenic contamination while exhibited apparent phytotoxicity and low arsenic uptake when grown in multiple metal/metalloid co-contaminated soils (Caille *et al.*, 2004).

To predict the performance and success of arsenic phytoextraction with *P. vittata*, a combined solubility-uptake model was established by Shelmerdine *et al.* (2009). According to the model, arsenic phytoextraction by *P. vittata* is only suitable for marginally contaminated sites with relatively high soil pH (>6.0). For example, for a heavily-contaminated soil (pH 5.6, total As concentration 1250 mg kg^{-1}, and labile As fraction of 3%), assuming an annual harvestable yield of 3 t ha^{-1}, no significant decrease in total soil arsenic content over 30 years is expected. Therefore, arsenic phytoextraction using *P. vittata* is applicable in lightly contaminated soils and can be used as a soil-polishing tool in combination with other conventional remediating strategies.

4.2.2 *Arsenic hyperaccumulation mechanisms*

Based on a number of studies, the unique mechanisms of As hyperaccumulation have been gradually unraveled, which appears to involve efficient As mobilization in the rhizosphere, rapid root uptake and enhanced frond translocation of As by *P. vittata* (Fig. 4.1).

Table 4.2. Typical greenhouse and field studies of arsenic phytoextraction by *P. vittata*.

Study site source of arsenic contamination	Initial soil arsenic [mg kg⁻¹]	Growth conditions	Soil properties	Study time	Frond arsenic [mg kg⁻¹ dw]	BF	Frond biomass (dw)	Reference
USA, Chromated copper arsenate (CCA)	190	Field site with subtropical climate	loamy, siliceous; pH 7.4–7.6	2 years	3186–4575	29–45	1.3 t ha^{-1} yr^{-1}	Kertulis-Tartar et al. (2006)
Australia, Arsenic-based pesticides	393–1903	Field site with subtropical climate	pH 4.8	5 months	775–2569	1–3	39.5 g fern^{-1}	Niazi et al. (2009)
England, As-rich parent material, mining, industrial activities, agrochemical application and biosolid disposal	8.8–3580	Pot trial in a greenhouse with day/night temperatures of 25/16°C and a 16 h photoperiod	pH 4.0–8.2 Cu 12.4–3,560 mg kg^{-1} Cd 69 mg kg^{-1} Pb 19–19,400 mg kg^{-1} Zn 57–21,000 mg kg^{-1}	9 months	9–3150	1.2–229	15.6 g fern^{-1} (2.5 t ha^{-1} yr^{-1} with planting density of 16 plants m^{-2})	Shelmerdine et al. (2009)
USA, Mining activities, arsenical insecticide, pesticide, CCA, and natural soil with high arsenic	23–640	Pot culture in a greenhouse	pH 6.7–7.9	16 months	110–6151	4.4–47.8	25–34 g fern^{-1}	Gonzaga et al. (2008)
USA, CCA	98	Pot culture in a greenhouse	pH 7.5	5 months	6000	3.0–87.2	18 g fern^{-1}	Tu et al. (2002)

BF is bioconcentration factor and is the ratio of total arsenic concentration in fronds to that in soil.

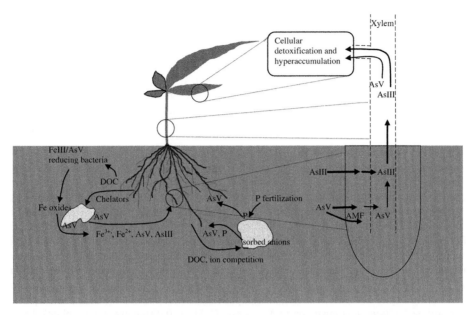

Figure 4.1.   Schematic diagram of rhizospheric mobilization, root uptake and frond translocation of arsenic in *P. vittata*. DOC, dissolved organic carbon and AMF, arbuscular mycorrhizal fungi.

4.2.2.1   *Arsenic mobilization via root exudates*

Both root exudates and bacteria associated with *P. vittata* have been shown to help arsenic solubilization in the rhizosphere. In a greenhouse study, *P. vittata* excreted 2 times more dissolved organic carbon (DOC) than a control non-hyperaccumulating fern *Nephrolepis exaltata* (Tu *et al.*, 2004b). As a result, the organic acids from root exudates of *P. vittata* induced 3–18 times higher mobilization of arsenic from insoluble arsenic minerals ($AlAsO_4$ and $FeAsO_4$) and an arsenic-contaminated soil as compared to *N. exaltata*. Besides root exudates, arsenic-resistant bacteria inhabiting *P. vittata* rhizosphere (*Pseudomonas* sp., *Comamonas* sp. and *Stenotrophomonas* sp.) have been shown to exhibit a remarkable ability to increase arsenic concentration in the uptake solution from $<5\,\mu g\,L^{-1}$ to 5.04–$7.37\,mg\,L^{-1}$ by solubilizing insoluble $FeAsO_4$ and $AlAsO_4$ (Ghosh *et al.*, 2011). The production of pyochelin-type siderophores by arsenic-resistant bacteria has been suggested to play a role in arsenic solubilization.

To further investigate the rhizosphere characteristics of *P. vittata* relevant for its use in phytoextraction and the effects of root uptake on arsenic redistribution in soils, a sequential extraction procedure has been developed to fractionate arsenic into five operationally-defined fractions with decreasing availability (Fitz and Wenzel, 2002). It includes non-specifically bound, specifically bound, bound to amorphous hydrous Fe/Al oxides, bound to crystalline hydrous Fe/Al oxides, and residual fractions. In comparison with non-hyperaccumulator *N. exaltata*, *P. vittata* was more efficient to access arsenic from all five fractions, leading to greater removal of arsenic (39–64% *vs*. 5–39%) from rhizosphere soils than *N. exaltata* (Gonzaga *et al.*, 2006). This observation seems related to 9% higher DOC and 0.5 unit higher pH in *P. vittata* rhizosphere, which tend to increase arsenic bioavailability by facilitating arsenic desorption from solid phase via anion competition. In addition, the majority of arsenic removed by *P. vittata* was from the major arsenic sink in soils (i.e., bound to Fe/Al hydrous oxides), accounting for 68% arsenic decrease in the rhizosphere. Fitz *et al.* (2003) reported that the difference in non-specifically bound arsenic (readily labile) between bulk and rhizosphere soils accounted for only 8.9% of total arsenic accumulated in *P. vittata*, again suggesting arsenic uptake was mainly from less available pools. Compared with the bulk soils, a 2–2.8-fold greater Fe and DOC concentration was observed in *P. vittata* rhizosphere soil, indicating that arsenic mobilization was mainly resulted from Fe solubilization by organic compounds in *P. vittata* rhizosphere. In the experiment with *P. vittata* growing in an arsenic-contaminated soil (total arsenic 2270 mg kg^{-1}) for 41 d, arsenic depletion and limited resupply in the root rhizosphere was successfully illustrated by a 19.3% decrease in arsenic flux from solid phase to soil solution using diffusive gradients in thin films (Fitz *et al.*, 2003). It seems DGT technique holds promise as an effective tool to monitor the bioavailability of target contaminant during and after phytoextraction.

4.2.2.2   *Efficient root uptake system*
In addition to being efficient in arsenic solubilization from the soil, *P. vittata* is also effective in arsenic uptake from soils. A comparison study between *P. vittata* and non-hyperaccumulating fern *N. exaltata* demonstrated that a more extensive root system and highly efficient root uptake contribute to the effective arsenic extraction by *P. vittata* (Gonzaga *et al.*, 2007b). Compared to *N. exaltata*, a 2.4–3.8 times greater root system in terms of biomass was developed by *P. vittata* after 8 weeks of growth in both arsenic-contaminated and control soils. Furthermore, arsenic root uptake efficiency, which is defined as arsenic accumulation in fronds or roots per unit root biomass was 8–23 times greater in *P. vittata* than that of *N. exaltata*, indicating a more efficient root uptake system in *P. vittata*. The more extensive root system coupled with much higher root uptake efficiency of *P. vittata* accounts, at least partially, for the 29-fold higher arsenic depletion from soil by *P. vittata* in comparison to *N. exaltata* (2.51 *vs.* 0.09 mg arsenic per plant). This is consistent with the study of Poynton *et al.* (2004) who reported a significantly lower Michaelis constant K_m for arsenate [As(V)] influx into the roots of *P. vittata* than *N. exaltata* (1.1–6.8 μM *vs.* 9.9–19.9 μM), indicating a higher affinity of transporter protein for As(V) in *P. vittata*.

As the major arsenic species in aerobic soils, As(V) shares P transport system in higher plants including *P. vittata* (Zhao *et al.*, 2009). As(V) uptake by *P. vittata* roots tend to be inhibited by increasing P in a competitive manner (Wang *et al.*, 2002). Nevertheless, in the presence of As(V) at 1–10 mg L^{-1} in the uptake solution, P concentration in *P. vittata* was increased by 6.3- and 2.2-fold in the roots (from 0.91 to 5.76 mg g^{-1}) and fronds (from 2.33 to 5.19 mg g^{-1}), respectively (Luongo and Ma, 2005), which is in direct contrast to other tested ferns with an average of 40% reduction in P level. Therefore, it is suggested that the maintenance of sufficient P in fern tissues via efficient root uptake in the presence of high arsenic may constitute an essential detoxification mechanism in *P. vittata*.

On the other hand, it should be noted that due to the restriction of fern root extension, arsenic phytoextraction using *P. vittata* from soil profiles beyond root zones is much slower (Fitz *et al.*, 2003). Therefore, it is necessary to evaluate fern rooting depth under field conditions, which will help to determine the effective depth of phytoextraction using *P. vittata*.

4.2.2.3   *Efficient arsenic translocation to fronds*
Not only does *P. vittata* have efficient root uptake system but also effective translocation mechanisms, making it the most striking attribute of an arsenic hyperaccumulator. Translocation factor (*TF*) has been used to characterize the effectiveness of plant arsenic translocation from the roots to fronds. As reported by Ma *et al.* (2001), arsenic *TF* in *P. vittata* reached 24 with frond arsenic being up to 7,234 mg kg^{-1} compared to root arsenic of 303 mg kg^{-1} after 20-week growth in an arsenic-contaminated soil (98 mg kg^{-1} As). Under variable soil arsenic concentrations from 6 to 1500 mg kg^{-1}, arsenic accumulation in *P. vittata* fronds increased rapidly to 755–15,861 mg kg^{-1} after two weeks with frond *BF* being 10.6–126 (Ma *et al.*, 2001). Following the first report unraveling the highly efficient translocation of arsenic in *P. vittata*, much research using hydroponics, greenhouse and field studies has consistently supported this constitutive trait of *P. vittata* (Natarajan *et al.*, 2009; Singh and Ma, 2006; Su *et al.*, 2008; Tu *et al.*, 2002). For instance, by growing accessions of *P. vittata* from both contaminated and uncontaminated environments in arsenic-contaminated soils (0–500 mg kg^{-1}), efficient arsenic translocation resulted in an average of *TF* value of 6.8 with frond *BF* being 11.7 to 21.6 (Zhao *et al.*, 2002).

The mechanisms responsible for the highly efficient arsenic translocation in *P. vittata* have been gradually unraveled. Arsenite [As(III)] is consistently present as the major species in *P. vittata* fronds with As(V) dominating the roots regardless arsenic species supplied (Kertulis *et al.*, 2005; Mathews *et al.*, 2010; Singh and Ma, 2006; Wang *et al.*, 2002).This suggests that efficient As(V) reduction and As(III) translocation as an important contributing factor to arsenic hyperaccumulation in *P. vittata*.

Su *et al.* (2008) reported that 93–98% of arsenic in the xylem sap of *P. vittata* was As(III) with either As(V) or As(III) being supplied, indicating a significantly higher mobility of As(III) than As(V) in xylem transport. Mathews *et al.* (2010) investigated the location of As(V) reduction by exposing *P. vittata* to 0.10 mM As(V) for 8 d. They found that As(III) concentrations increased significantly from 7% in the roots up to 68% and 71% in the rhizome sap and rhizome tissue. Along with upward translocation to the fronds, 86% As(III) in the frond sap and 90–100% As(III) in the pinnae were recorded, indicating a remarkable reduction capacity of the fronds as well as the rhizome. This is supported by Tu *et al.* (2004c) who has shown that upon exposing excised tissues of *P. vittata* to 0.67 mM As(V) for 2 d, 86% and 24% As(III) was present in the excised pinnate and roots. Taken together, it is conceivable that rhizomes and fronds are primarily responsible for As(V) reduction in *P. vittata* with increasingly more reduction along upward translocation. Besides, little arsenic efflux from the roots to external media and lack of a strong arsenic sequestration in roots both facilitate the highly efficient arsenic translocation in *P. vittata* (Su *et al.*, 2008; Zhao *et al.*, 2009).

Arsenic phytoextraction with *P. vittata* has been demonstrated to be affected by plant age, with young ferns exhibiting higher arsenic translocation than the older ones (Gonzaga *et al.*, 2007a; Santos *et al.*, 2008; Tu *et al.*, 2004a). For example, 36% more arsenic was accumulated in 2-month old *P. vittata* than 4–16 month old plants after 8-week growth in an arsenic-contaminated soil, which is likely associated with the higher metabolic activity in younger plants (Gonzaga *et al.*, 2007a). In addition, arsenic *TF* was reduced from 3.2 to 1.6 from old to young ferns, indicating decreased arsenic translocating ability of *P. vittata* with increasing plant age. Furthermore, frond biomass after 8 weeks growth increased by 39, 6.9, 2.0 and 1.1 times for *P. vittata* of increasing age of 2-, 4-, 10- and 16-month old, respectively. These results highlight the necessity to use young ferns in phytoextraction and harvest fronds before they senesce to minimize the remediating time and potential arsenic being leached from senesced fronds by rainwater in the field.

4.2.3 *Potential improvement*

4.2.3.1 *Phosphorous amendment*

To achieve more effective phytoextraction, it is essential to employ proper agronomic techniques and plant management such as fertilizing and rhizosphere manipulation. One key strategy towards efficient phytoextraction is to enhance arsenic availability in soils. P amendment can be used in assisting arsenic uptake by *P. vittata* in the presence of toxic metals (e.g., Pb, Zn, and Cd) through increasing plant biomass and soil arsenic bioavailability via competitive anion exchange together with reduced metal toxicity by immobilization (Cao *et al.*, 2003; Fayiga and Ma, 2006; Tu and Ma, 2003). A maximum of P/As ratio of 1.2 in soil solution or 1.0 in the fronds has been suggested for an improved performance of *P. vittata* in arsenic phytoextraction by enhancing fern biomass and arsenic uptake (Tu and Ma, 2003). For instance, with a range of P added to the tested soil containing 400 mg As(V) kg^{-1}, a maximum of 26% soil arsenic extraction by *P. vittata* was recorded at water soluble P/As molar ratio of 2.3 after 20 weeks growth in a greenhouse experiment. Similarly, for another arsenic hyperaccumulator *Pityrogramma calomelanos* discovered in Thailand, a significant increment of frond arsenic content of 14 mg plant^{-1} was reported after 8-week growth in the field containing 136–269 mg As kg^{-1} with the addition of 100 mg P kg^{-1} soil (Jankong *et al.*, 2007). In hydroponics, split P addition ($134 + 66$ µM) during *P. vittata* acclimation and after arsenic exposure (145 µg L^{-1}) has been shown to induce 1.5-folder higher efficiency in frond As accumulation in the younger ferns (45-d-old) compared to the older ones (90-d and 180-d-old) (Gonzaga *et al.*, 2008), suggesting the more efficient stimulation of P on As uptake in the young ferns. However, in addition to increasing As availability in soils, P also competes with As for plant uptake so caution needs to be excised when adding P so plant arsenic uptake is not adversely affected.

4.2.3.2 *Mycorrhizal symbiosis*

Regarding rhizosphere manipulation via mycorrhizal symbiosis, which has a well-documented role in improving P uptake, it can increase arsenic accumulation and frond biomass in *P. vittata* (Agely et al., 2005; Fitz and Wenzel, 2002). *Glomus microaggregatum, G. mosseae, G. brohultii* and *G. geosporum* represent the most common arbuscular mycorrhizal fungi (AMF) in the rhizosphere of *P. vittata* (Wu et al., 2009) At the exposure of 3.8–75 mg L^{-1} As(V) in hydroponics, inoculation of an uncontaminated isolate of *G. mosseae* almost doubled arsenic influx into *P. vittata* roots relative to the control (25–105 *vs.* 14–55 mg As kg^{-1} dw h^{-1}) during 20-min uptake experiment (Wu et al., 2009). In a greenhouse experiment with total soil arsenic of 100 mg kg^{-1}, 2–5 times more arsenic accumulation in the fronds was found in *P. vittata* colonized by a community of AMF, with the enhancement being more evident with increasing soil P level as compared to those without colonization (Agely et al., 2005). Considering the fact that arsenic acts as a chemical analog of P and the well documented role of AMF in enhancing P nutrition for the host plants, it is not surprising to find enhanced arsenic accumulation by mycorrhizal *P. vittata* with concurrent increase of plant biomass. However, the contribution of AMF to plant growth and arsenic translocation in *P. vittata* has been shown to largely depend on the species of AMF (Trotta et al., 2006), implying arsenic phytoextraction with *P. vittata* could be optimized by selected AMF symbiosis.

4.2.4 *Potential environmental risks*

4.2.4.1 *Invasive risk*

As a hardy and perennial fern species, *P. vittata* propagates quickly and is easy to maintain in a humid tropic/subtropic climate, which facilitates arsenic phytoextraction using *P. vittata*. However, from an ecological point of view, there is a concern regarding its invasive potential considering it is classified as a type-II invasive species in Florida. It is important to employ young ferns (e.g., 2 month-old), which exhibits higher capacity in arsenic accumulation than the old ferns (e.g., 16 month-old) (Gonzaga et al., 2007a). Such practice can reduce the potential of fern invasion with little spores being produced during phytoextraction. In addition, to reduce its invasion risk and overcome the geographical limitation of arsenic hyperaccumulators, exploration of indigenous species with potential for arsenic phytoextraction from local sites rich in arsenic, e.g., mining areas, may provide alternative options, which are well-adapted to the local environment (Antosiewicz et al., 2008; Mahmud et al., 2008; Visoottiviseth et al., 2002).

4.2.4.2 *Disposal of arsenic-rich biomass*

Safe management and economical disposal of arsenic-loaded biomass remains unsolved. Improper treatment of arsenic-rich plant materials may pose threats to ecosystem safety. For *P. vittata*, inorganic As(III) accounts for ~94% of the total arsenic in the fronds after 18 weeks growth in soil containing 50 mg As kg^{-1} (Tu et al., 2003). Furthermore, when the fronds were air-dried, the amount of leached arsenic substantially increased, with arsenic concentration in the leachate reaching 0.65 mg L^{-1} after 5 d of drying. The same holds true for arsenic hyperaccumulating fern *P. calomelanos* which grows naturally on arsenic-contaminated sites in southern Thailand, arsenic in the fronds is primarily present as inorganic species with 86–93% being water-extractable with the majority being As(III) (60–72%) (Francesconi et al., 2002). Taken together, these results suggest that arsenic concentrated in the hyperaccumulating ferns has high water solubility and hence the arsenic-rich biomass should be properly managed and kept away from water supply to minimize secondary contamination.

As for the possible disposable options, there is a paucity of data regarding the post-treatment technology and associated cost-benefit analysis. To effectively reduce plant biomass, incineration has been considered and more than 90% of plant biomass can be reduced by this means

(Sas-Nowosielska *et al.*, 2004). As a volatile metalloid, arsenic tends to be easily released by vaporization during thermal processing. Recent study has revealed that 24% of the accumulated arsenic in *P. vittata* could be emitted during incineration at 800°C with a 94% loss of biomass (Yan *et al.*, 2008). The residual ash provides an enriched and recoverable ore with arsenic concentration being increased by 11-fold. While the development of a highly efficient capture system targeting arsenic-containing flue gas seems essential for further application of this technology.

In addition, water extraction of the harvested biomass may be feasible considering its high water solubility of arsenic. The resulting arsenic-rich extract could be recycled by an industry. Furthermore, decreased arsenic level in plant biomass could make the residue non-hazardous waste, with lower fee in landfill disposal. However, there is still large uncertainty for this option regarding arsenic extraction efficiency, residual arsenic concentration, and economic effectiveness. Alternatively, high-arsenic biomass could be disposed in marine system (Francesconi *et al.*, 2002), which has a high detoxification capacity by transforming inorganic arsenic into essentially non-toxic organic forms by natural biochemical processes. While preliminary test and thorough assessment with regard to the ecological effect of this scenario is needed to minimize potential environmental risks. In addition, recent study has shown polymeric encapsulation technique could efficiently stabilize arsenic-bearing solid residual (e.g., iron-based sorbent) producing more than an order of magnitude lower concentration of leached arsenic than the conventional cement encapsulation (Shaw *et al.*, 2008). However, the applicability of polymeric matrices to encapsulate arsenic-rich biomass needs both standard and landfill simulation leaching tests. In brief, there is a long way to go to establish a technically feasible, economically acceptable and environmentally safe disposal method.

4.3 PHYTOSTABILIZATION

4.3.1 *Indigenous tolerant species with low* TF

In sites contaminated with high-level of arsenic and other toxic metals (e.g., Cu, Zn, and Cd), phytostabilization is advantageous by immobilizing arsenic and reducing its exposure risks to ecological receptors (Table 4.3). High adaptability of spontaneous species with multiple metal/metalloid tolerance has been widely reported to serve as potential candidates for phytostabilization of the contaminated sites (Antosiewicz *et al.*, 2008; Vamerali *et al.*, 2009; Whiting *et al.*, 2004). For instance, in a recent phytoremediation trial on a contaminated site with 0.15 m layer of gravelly soil over a 0.7 m layer of cinders containing As, Co, Cu, Pb and Zn, 100% survival of indigenous *Populus* and *Salix* were obtained after two-year of experiments with soil amelioration (e.g., mixing with imported soils, tillage, and fertilization) in spite of 16–92% lower tissue biomass than the control (Vamerali *et al.*, 2009). The fact that trace metals were preferentially accumulated in the woody roots (84–89%) with marginal shoot translocation suggests the potential utilization of native *Populus* and *Salix* in phytostabilization of arsenic-contaminated sites.

4.3.2 *Substrate improvement by legumes*

To overcome the general poor nutritional status in degraded sites with arsenic contamination, legume plants with good N fixation capacity and strong root system serve as promising pioneering colonizer species for substrate improvement and revegetation. Furthermore, most species of legume generally exhibited low capability for shoot translocation of metals and hence pose low risks of exposure to other organisms in the food chain. For example, by growing white lupin in soils contaminated by the spill of acid pyrite sludge with co-occurrence of As and Cd, symbiosis with rhizobia were successfully established by lupin plants with root nodule formation, although the efficiency of N-fixation were reduced by 30–40% due to metal toxicity and inhospitable substrate. Simultaneously, soil acidity was mitigated with a marked increase of soil pH from

Table 4.3. Typical studies of arsenic phytostabilization.

Location and contaminant source	Contaminant concentration [mg kg^{-1}]	Soil amelioration	Plant species	Remediating time	Outcome	Reference
Torviscosa, Udine, Italy Cinder waste from pyrite ore roasting	As 886 Co 100 Cu 1735 Pb 493 Zn 2404	Metal-rich cinders covered with an unpolluted 0.15 m layer of soil. Both cinders and soil were mixed with sand (1:1 w/w) in pot experiment.	*Populus and Salix*	2 years	Coarse and fine roots provided a significant sink for contaminants with both TF and BCF <1	Vamerali *et al.* (2009)
Rixton clay pits Kidsgrove Merton Bank, UK Industrial waste	As 60–78 Cu ~508 Cd ~36	Green waste compost (GWC, 30% v/v) Biochar (20% v/v)	*Miscanthus*	8 months	Biomass yield increased by 2–4 times with GWC; Both TF and BCF <1	Hartley *et al.* (2009)
Aznalcóllar, Spain; Mine tailing spill	As 49–339 Pb 73–607 Sb 4.5–37.7	Organic matter Ca-rich amendments	*Quercus ilex* subsp. *Olea europaea* *Populus alba* L. *Phillyrea angustifolia* L. *Pistacia lentiscus* L. *Rosmarinus officinalis* L., *Retama sphaerocarpa* L. *Tamarix africana* Poir.	since 1999	BCF <0.03 for As, Pb, Sb in most species; BCF for Cd in *Populus alba* approaching 2	Domínguez *et al.* (2008)
Sugar Brook, UK Fazakerley, UK Kirby Moss, UK Merton Bank, UK Cromdale Grove, UK; Landfill, industrial waste, sewage sludge	As 4.9–5266 Zn 2.8–1300 Cu 10–880 Ni 10–109 Pb 45–1770	Soil turning to a depth of 30 cm followed by rotovation and weed-control treatment	*Salix* *Populus* hybrids *Alnus* *Betula* *Larix*	3 years	Low mobility and plant transfer for As, Pb, Cu, Ni with BCF <1; Higher transfer of Cd and Zn with BCF up to 4–13 in *Salix*	French *et al.* (2006)
Sanlúcar la Mayor, Spain Mine tailing spill	Available Cd 0.49 μM Available As 40 μM		*Lupinus albus* (white lupin)	3 weeks for pot study; 1–6 months for field trial	Both TFs of Cd and As < 1; Decreased soluble Cd and As in planted soil	Vázquez *et al.* (2006)

3.9 to 6.5 after two weeks growth of white lupin in a pot experiment. Furthermore, soil soluble concentrations of As and Cd were reduced by 56% and 86% after five-month growth in field conditions with the roots being the major sink of As and Cd (Vazquez *et al.*, 2006).

4.3.3 *Fe oxides and biochar*

In some cases, amendments such as Fe oxides are essential to mitigate arsenic toxicity and hence facilitate plant survival. For example, two highly-contaminated mine tailings in South Korea contained arsenic levels up to 6670 and 56,600 mg kg^{-1}, resulting in high arsenic toxicity to plants. Reduction of $\sim$70–80% available arsenic was achieved by adding amorphous Fe with the majority of arsenic bond to the stable Fe precipitates (Kim *et al.*, 2003). In a field trial involving four arsenic-contaminated sites from UK containing 748 mg As kg^{-1}, a mean reduction of 22–32% arsenic uptake by tested vegetables was recorded upon the addition of 0.2–0.5% Fe as ferrous sulfate in the top soils (Warren *et al.*, 2003), confirming the high capacity of iron in soil arsenic immobilization. In a field trial of a grassland established on an abandoned chemical waste site, soil amendment with Fe(III) plus lime was the most efficient treatment with the labile arsenic being reduced by 92%. An 8% reduction in leached arsenic upon the application of lime was observed, resulting from the binding of arsenic with Ca^{2+} and resulting in reduced As mobility (Hartley *et al.*, 2009a).

Biochar, as a promising soil amendment, has important environment implications for the biochemical behaviors of metals in soils. Regardless the type of feedstock and pyrolysis conditions, biochar with relatively high cation exchange capacity consistently shows adsorption capacity towards metal cations but little binding ability for arsenic species at typical environmental pH (Mohan *et al.*, 2007). It should be noted that arsenic liability in soils increased to varying extent upon the application of biochar with higher arsenic concentrations in water-soluble and surface-adsorbed pool, probably due to raised soil pH by biochar (Hartley *et al.*, 2009b; Namgay *et al.*, 2010). For example, in a pot experiment with maize, the extractable arsenic in soil increased from 5.16 mg kg^{-1} in the control soil to 5.96 mg kg^{-1} in the biochar treatment (15 g biochar kg^{-1} soil) (Namgay *et al.*, 2010). Similarly, in a 60-d field experiment, arsenic concentration in soil pore water increased by 30 fold with biochar application (soil:biochar = 2:1, v/v). These results highlight the potential risk of arsenic mobilization upon biochar application to arsenic-containing soils. To effectively immobilize arsenic upon the application of biochar, iron oxides can be employed considering their high affinity to arsenic (Hartley and Lepp, 2008). Therefore, the combination of biochar and iron oxides can be a useful strategy to immobilize arsenic while improving soil fertility.

4.3.4 *Phosphate*

As a chemical analog of As(V), P is an effective competitor of As(V) for binding sites in soils. Due to the competitive anion exchange, increased bioavailability and plant uptake of arsenic has been well-documented upon P application (Cao and Ma, 2004; Hartley *et al.*, 2009a). In a pot experiment with soil arsenic at 0, 15, and 30 mg kg^{-1}, elevated arsenic accumulation in both rice grain and straw with lower grain yield was observed after P application of 50 mg kg^{-1} P (Hossain *et al.*, 2009). For example, arsenic concentrations in rice grain and straw increased from 0.64 and 5.77 mg kg^{-1} to 0.71 and 6.21 mg kg^{-1} upon P addition at 30 mg kg^{-1} arsenic. As a result, the grain yield (g pot^{-1}) was reduced by 33–66%. Furthermore, arsenic concentration associated with Fe plaque was reduced by 20% on average, indicating higher arsenic solubility induced by P in rice rhizosphere. This highlights the risk that P fertilization may induce arsenic mobilization in soils and increase its uptake by food crops. Similar case has been reported during phytostabilization of a gold mine tailings with elevated arsenic ($>$1000 mg kg^{-1}) in New Zealand, where P has been employed to improve the hostile environment for better plant establishment (Mains *et al.*, 2006). As a result, the leached arsenic increased proportionally to the amount of P applied. With P amendment at 3 g m^{-2}, up to 0.5 and 0.9 mg L^{-1} arsenic was leached from the bare and planted lysimeters during the five-month field trial. Furthermore, an interactive effect of plant species and type of P amendment on arsenic mobilization has been observed (Mains *et al.*, 2006). For example, organic bioboost amendment (dehydrated sewage sludge) induced higher leached arsenic than superphosphate fertilizer from barley (2.4 *vs.* 1.0 mg L^{-1}), but not for rye, corn or blue lupin. The fact that soil arsenic mobilization may occur upon plant establishment with P amendment highlights the necessity of pre-assessment to choose proper amendment strategy and plant candidate to minimize arsenic leaching and soil-plant transfer during phytostabilization.

4.3.5 *Organic matter*

As a complex mixture of varying components, organic matter (OM) has inconsistent effects on arsenic mobility, mainly due to the type of compost applied, degree of humification and pH variation (Juwarkar *et al.*, 2008; Lagerkvist *et al.*, 2008; Shiralipour, 2002). For example, a significant increase in soil soluble arsenic from 5.7 mg L^{-1} to 7.1 mg L^{-1} was reported upon application of municipal solid waste and biosolids compost in a greenhouse experiment (soil pH 6.87) (Cao and Ma, 2004). Increased OM under neutral soil pH may improve microbial activity, and hence facilitate As(V) reduction to more mobile As(III). At the end of the experiment, As(III) concentration in soil solution was ∼20–24% in OM treatment compared to <10% in the control. In contrast, a reduction of water-soluble arsenic from 32 mg L^{-1} to 25 mg L^{-1} upon OM application was obtained in an arsenic-spiked soil with pH of 5.45 in the same experiment (Cao and Ma, 2004). This is probably due to arsenic adsorption onto OM in acidic conditions. Therefore, it is critical to monitor arsenic availability using OM to reduce the potential environmental risks of arsenic mobilization while improving substrate conditions.

4.3.6 *Mycorrhiza*

Proper inoculates of As-tolerant mycorrhiza can serve as a potential strategy to confer host-enhanced tolerance and facilitate phytostabilization by selectively accumulating P over As(V) (Sharples *et al.*, 2000). For example, short-term As(V) and P uptake kinetics of mycorrhizal fungus *Hymenoscyphus ericae* from the roots of *Calluna vulgaris* were identical for the mine-impacted and control population. However, enhanced efflux of As(III) from mycelia preloaded with 0.1 mM As(V) for 1 h was observed in the resistant *H. ericae* from the mine population (14.4% h^{-1} *vs.* 6.6% h^{-1}), indicating the role of fungus as biofilter to maintain low plant arsenic via efficient As(III) efflux (Sharples *et al.*, 2000). White clover (*Trifolium repens* Linn.) and ryegrass (*Lolium perenne* L.), which represent legumes and grasses that are commonly used in revegetation, exhibit high dependence on the mycorrhizal associations with *G. mosseae* for surviving in soils with severe arsenic contamination (Dong *et al.*, 2008). Using a compartmented cultivation system, mycorrhizal inoculation of *G. mosseae* increased plant P concentration by 50–200%, with shoot arsenic being reduced by 9%–30% in the presence of 1 and 205 mg kg^{-1} arsenic. The data indicate the beneficial effect and potential use of AMF in phytostablization via selectively accumulating P over As(V).

4.4   PHYTOEXCLUSION

For agriculture soils contaminated by arsenic, it is impractical to employ non-food crops for either phytoextraction or phytostabilization. Rice, which feeds half of the world's population, is more efficient in arsenic accumulation than other cereals through efficient silicon transport pathway (Ma *et al.*, 2008), particularly under flooding conditions. As a result, elevated arsenic has been widely reported in paddy rice with arsenic *TF* often approaching unity (Abedin *et al.*, 2002; Stroud *et al.*, 2010; Voegelin *et al.*, 2010; Williams *et al.*, 2007; 2009). To improve agriculture sustainability and food safety, a range of agronomic strategies and biotechnologies have been developed (Zhao *et al.*, 2010), which may provide effective solutions to remediate arsenic-contaminated agriculture soils and reduce arsenic uptake by rice.

4.4.1 *Water management*

Because arsenic mobility and toxicity in paddy soils is largely controlled by soil redox potential, water management can be effective in reducing arsenic mobilization resulting from reductive dissolution of Fe hydroxides in anaerobic conditions. Compared with the conventional flooding cultivation, arsenic availability and uptake by rice can be remarkably decreased under aerobic conditions (Li *et al.*, 2009; Xu *et al.*, 2008). For example, compared with control, 4–16 times higher soluble arsenic concentrations were reported with As(III) accounting for 81–95% of the arsenic in flooding treatment. With efficient As(III) uptake system in rice, the enhanced arsenic solubility in flooded soils caused 10–15 fold higher grain arsenic concentrations than those under the aerobic treatment (Xu *et al.*, 2008). Similarly, in a field trial with paired plots to compare raised bed and conventional flooding, a significantly higher redox potential was observed in the raised bed as compared with paddy conditions (0 *vs.* 120 mV in 0–15 cm soil), resulting in 3–6 fold lower arsenic concentrations in rice straw under aerobic condition than those in the flooded treatment (Duxbury *et al.*, 2007). However, in the presence of Cd, aerobic treatment tends to mobilize soil Cd through the oxidation of CdS into $CdSO_4$ (Kawasaki *et al.*, 2009), indicating the potential negative effect of aerobic rice cultivation in soils contaminated with both As and Cd.

4.4.2 *Silicon fertilization*

Efficient As(III) uptake by rice is via Si transport system (Ma *et al.*, 2008), suggesting enhanced Si availability can mitigate arsenic transfer in soil-rice system as well as improving grain yield (Ma and Yamaji, 2006). In a pot experiment, Si fertilization (20 g SiO_2 kg^{-1} soil) decreased arsenic concentration in rice straw and grain by 78% and 16%, in spite of the 1.5–2 fold higher arsenic concentration in soil solution [with 78–100% As(III)] (Li *et al.*, 2009). However, increased As availability did not translate to higher rice arsenic uptake probably due to enhanced competition of silicic acid for As(III) by plant uptake. Moreover, Si fertilization affected arsenic fraction by reducing inorganic arsenic level while enhancing arsenic methylation in both rice grain and husk. For example, 59% reduction of inorganic arsenic concentration and a concurrent 33% increase in dimethylarsinic acid concentration was found in rice grain with Si application (Li *et al.*, 2009). Therefore, Si fertilization is a promising strategy to reduce arsenic uptake and phytotoxicity in arsenic-contaminated soil-rice system.

4.4.3 *Arsenic sequestration by Fe plaque*

For paddy rice as well as other aquatic species, iron plaque formed on the root surface due to oxygenation of rhizosphere exhibits high capacity for retaining As(V), and therefore could effectively diminish arsenic influx into rice roots. In a pot culture experiment with rice growing under flooded condition, arsenic concentrations in the rhizosphere soil solutions were remarkably decreased, being 2.5-fold and 16-fold lower upon the amendment of amorphous iron at 0.1 and 0.5%, resulting from enhanced arsenic sequestration by Fe plaque. In comparison with control, arsenic content binding to Fe-plaque was 3–4 times higher in Fe treatments (0.51 *vs.* 1.49 and 2.41 mg pot^{-1}). As a result, arsenic concentrations in rice shoots were significantly reduced, accounting for only 1/7–1/2 that in the control (Ultra *et al.*, 2009).

Recent studies have further demonstrated significant variation in Fe plaque formation and arsenic sequestration among rice cultivars (Mei *et al.*, 2009) and genotypes (Liu *et al.*, 2006). Based on the pot experiment with 25 rice cultivars, significant negative correlation ($p < 0.001$) was observed between grain arsenic and root porosity and the rate of radial O_2 loss (Mei *et al.*, 2009). Rice cultivars with higher root porosity and rate of radial O_2 loss exhibit higher oxidizing ability by releasing more O_2 to the rhizosphere, and hence possess higher capacity in limiting arsenic influx into rice roots via effective arsenic fixation by Fe plaque. Taken together, Fe amendment and breeding provide potential strategy to minimize arsenic transport into rice by enhancing arsenic binding to the rhizosphere Fe plaque.

4.4.4   Pretreatment of arsenic-contaminated irrigating water

In South and Southeast Asia, extensive irrigation with arsenic-contaminated groundwater accounts predominantly for the elevated arsenic in paddy soils. Besides the above strategies to minimize arsenic contamination in soil-rice system, it is equally important to take proper measures to scavenge arsenic from irrigation water. Arsenic phytofiltration, an emerging technology, has been tested in both laboratory- and pilot-scale experiments using arsenic hyperaccumulators such as *P. vittata*. With initial As(V) concentration ranging from 20 to $200 \, \mu g \, L^{-1}$, both *P. vittata* and another arsenic hyperaccumulator *P. cretica* exhibited high removal efficiency to reduce arsenic to below the drinking water limit of $10 \, \mu g \, L^{-1}$ within 24 h (Huang *et al.*, 2004). Furthermore, rapid arsenic removal has also been observed by repeatedly using the same batch of *P. vittata* after a short recovery (12–36 h), indicating the high sustainability of this technology. While in the presence of $1.6 \, mg \, L^{-1}$ P with initial arsenic of $200 \, \mu g \, L^{-1}$, a short term (60 min) arsenic influx into *P. cretica* roots was decreased by 88% as a result of the strong anion competition for root uptake (Huang *et al.*, 2004). However, for the well-established *P. vittata* (6–7 month old), P treatment (1.2 and $2.4 \, mg \, L^{-1}$) exerted little effect on arsenic depletion, with arsenic in the contaminated groundwater decreased from initial $130 \, \mu g \, L^{-1}$ to less than $10 \, \mu g \, L^{-1}$ in 2 d (Natarajan *et al.*, 2009).

In a pilot scale phytofiltration system, arsenic concentration in the outflow was invariably less than $2 \, \mu g \, L^{-1}$ through 84 d demonstration with initial arsenic between 6.6 and $14 \, \mu g \, L^{-1}$ and flow rate between 255 and $1900 \, L \, d^{-1}$. Besides, arsenic removal efficiency was unaffected by day length, light intensity and humidity (Elless *et al.*, 2005), demonstrating the high reliability of this technology. Therefore, to reduce arsenic buildup in agriculture soils, phytofiltration is potentially useful for arsenic depletion from irrigation water in the regions where prevalent arsenic contamination occurs in water supply. However, phytofiltration with *P. vittata* is limited to subtropical and tropical regions and more suitable for small-scale water treatment considering the increased cost of fern maintenance in cold areas and huge space need for large-scale treatment system.

4.5   CONCLUSIONS

Arsenic phytoextraction with hyperaccumulators such as *P. vittata* has a potential for soil remediation with low to moderate arsenic contamination. Arsenic phytoextraction technologies based on greenhouse and field studies mainly include candidate plant screening, fertilizing, rhizosphere manipulation, growth timing, and harvest method, which provide essential basis for larger scale application of this technology. Proper measures need to be taken to diminish fern invasive risks particularly in eco-fragile regions and to achieve safe disposal of arsenic-rich biomass.

Arsenic phytostabilization with indigenous tolerant species with low translocation capacity is advantageous for heavily contaminated sites with high levels of arsenic and other toxic metals. Amendments such as iron oxides, phosphate, organic matter, N-fixation legume and mycorrhizal inoculation are important for plant survival in hostile environment and serve as core strategies to facilitate the success of phytostabilization.

To alleviate arsenic contamination in soil-rice system, a range of agronomic strategies and biotechnologies from water management and Si and Fe fertilization to pretreatment of irrigation water provide effective phytoexclusion measures to remediate arsenic-contaminated agriculture soils and reduce arsenic uptake by rice.

ACKNOWLEDGEMENT

This project is supported by the Construct Program of the Key Discipline in Hunan Province, China, Hunan Provincial Natural Science Foundation of China (No.13JJ4044), and program for excellent talents in Hunan Normal University (No.ET12405).

REFERENCES

Abedin, M. J., Cresser, M. S., Meharg A.A., Feldmann, J. & Howells, J.C.: Arsenic accumulation and metabolism in rice (*Oryza sativa* L.). *Environ. Sci. Technol.* 36 (2002), pp. 962–968.

Agely, A.A., Sylvia, D.M. & Ma, L.Q.: Mycorrhizae increase arsenic uptake by the hyperaccumulator Chinese Brake fern (*Pteris vittata* L.). *J. Environ. Qual.* 34 (2005), pp. 2181–2186.

Ali, M.A., Badruzzaman, A.B.M., Jalil, M.A., Hossain, M.D. & Ahmed, M.F.: Fate of arsenic extracted with groundwater. In: M.F. Ahmed (ed): *Fate of arsenic in the environment*. Dhaka: ITN Int. 2003, pp. 7–20.

Antosiewicz, D.M., Escude-Duran, C., Wierzbowska, E. & Sklodowska, A.: Indigenous plant species with the potential for the phytoremediation of arsenic and metals contaminated soil. *Water Air Soil Pollut.* 193 (2008), pp. 197–210.

Arai, Y., Lanzirotti, A., Sutton, S., Davis, J.A. & Sparks, D.L.: Arsenic speciation and reactivity in poultry litter. *Environ. Sci. Technol.* 37 (2003), pp. 4083–4090.

Brammer, H. & Ravenscroft, P.: Arsenic in groundwater: A threat to sustainable agriculture in South and South-east Asia. *Environ. Int.* 35 (2009), pp. 647–654.

Caille, N., Swanwick, S., Zhao, F.J. & McGrath, S.P.: Arsenic hyperaccumulation by *Pteris vittata* from arsenic contaminated soils and the effect of liming and phosphate fertilisation. *Environ. Pollut.* 132 (2004), pp. 113–120.

Cao, X.D. & Ma, L.Q.: Effects of compost and phosphate on plant arsenic accumulation from soils near pressure-treated wood. *Environ. Pollut.* 132 (2004), pp. 435–442.

Cao, X., Ma, L.Q. & Shiralipour, A.: Effects of compost and phosphate amendments on arsenic mobility in soils and arsenic uptake by the hyperaccumulator, *Pteris vittata* L. *Environ. Pollut.* 126 (2003), pp. 157–167.

Chirenjea, T., Ma, L.Q., Chen, M. & Zilliouxb, E.J.: Comparison between background concentrations of arsenic in urban and non-urban areas of Florida. *Adv. Environ. Res.* 8 (2003), pp. 137–146.

Dittmar, J., Voegelin, A., Maurer, F., Roberts, L.C., Hug, S.J., Saha, G.C., Ali, M.A., Badruzzaman, A.B.M. & Kretzschmar, R.: Arsenic in soil and irrigation water affects arsenic uptake by rice: complementary insights from field and pot studies. *Environ. Sci. Technol.* 44 (2010), pp. 8842–8848.

Dominguez, M.T., Maranon, T., Murillo, J.M., Schulin, R. & Robinson, B.H.: Trace element accumulation in woody plants of the Guadiamar Valley, SW Spain: A large-scale phytomanagement case study. *Environ. Pollut.* 152 (2008), pp. 50–59.

Dong, Y., Zhu, Y.G., Smith, F.A., Wang, Y.S. & Chen, B.D.: Arbuscular mycorrhiza enhanced arsenic resistance of both white clover (*Trifolium repens* Linn.) and ryegrass (*Lolium perenne* L.) plants in an arsenic-contaminated soil. *Environ. Pollut.* 155 (2008), pp. 174–181.

Duxbury, J.M.: Remediation of arsenic for agriculture sustainability, food security and health in Bangladesh; FAO Water Working Paper; FAO, Rome, Italy, 2007, p. 18.

Elless, M.P., Poynton, C.Y., Willms, C.A., Doyle, M.P., Lopez, A.C., Sokkaryb, D.A., Fergusona, B.W. & Blaylock, M.J.: Pilot-scale demonstration of phytofiltration for treatment of arsenic in New Mexico drinking water. *Water Res.* 39 (2005), pp. 3863–3872.

Fayiga, A.O. & Ma, L.Q.: Using phosphate rock to immobilize metals in soil and increase arsenic uptake by hyperaccumulator *Pteris vittata*. *Sci. Total Environ.* 359 (2006), pp. 17–25.

Fitz, W.J. & Wenzel, W.W.: Arsenic transformations in the soil-rhizosphere-plant system: fundamentals and potential application to phytoremediation. *J. Biotechnol.* 99 (2002), pp. 259–278.

Fitz, W.J., Wenzel, W.W., Zhang, H., Nurmi, J., Stipek, K., Fischerova, Z., Schweiger, P., Kollensperger, G., Ma, L.Q. & Stingeder, G.: Rhizosphere characteristics of the arsenic hyperaccumulator *Pteris vittata* L. and monitoring of phytoremoval efficiency. *Environ. Sci. Technol.* 37 (2003), pp. 5008–5014.

Francesconi, K., Visoottiviseth, P., Sridokchan, W. & Goessler, W.: Arsenic species in an arsenic hyperaccumulating fern, *Pityrogramma calomelanos*: a potential phytoremediator of arsenic-contaminated soils. *Sci. Total Environ.* 284 (2002), pp. 27–35.

French, C.J., Dickinson, N.M. & Putwain, P.D.: Woody biomass phytoremediation of contaminated brownfield land. *Environ. Pollut.* 141 (2006), pp. 387–395.

Gonzaga, M.I.S., Santos, J.A.G. & Ma, L.Q.: Arsenic chemistry in the rhizosphere of *Pteris vittata* L. and *Nephrolepis exaltata* L. *Environ. Pollut.* 143 (2006), pp. 254–260.

Gonzaga, M.I.S., Ma, L.Q. & Santos, J.A.G.: Effects of plant age on arsenic hyperaccumulation by *Pteris vittata* L. *Water Air Soil Poll.* 186 (2007a), pp. 289–295.

Gonzaga, M.I.S., Santos, J.A.G., Comerford, N.B. & Ma, L.Q.: Comparison of root-system efficiency and arsenic uptake of two fern species. *Commun. Soil Sci. Plant Anal.* 38 (2007b), pp. 1163–1177.

Gonzaga, M.I.S., Santos, J.A.G. & Ma, L.Q.: Phytoextraction by arsenic hyperaccumulator *Pteris vittata* L. from six arsenic-contaminated soils: repeated harvests and arsenic redistribution. *Environ. Pollut.* 154 (2008), pp. 212–218.

Ghosh, P., Rathinasabapathi, B. & Ma, L.Q.: Arsenic-resistant bacteria solubilized arsenic in the growth media and increased growth of arsenic hyperaccumulator *Pteris vittata* L. *Bioresour. Technol.* 102 (2011), pp. 8756–8761.

Hartley, W. & Lepp, N.W.: Remediation of arsenic contaminated soils by iron-oxide application, evaluated in terms of plant productivity, arsenic and phytotoxic metal uptake. *Sci. Total Environ.* 390 (2008), pp. 35–44.

Hartley, W., Dickinson, N.M., Clemente, R., French, C., Piearce, T.G., Sparke, S. & Lepp, N.W.: Arsenic stability and mobilization in soil at an amenity grassland overlying chemical waste (St. Helens, UK). *Environ. Pollut.* 157 (2009a), pp. 847–856.

Hartley, W., Dickinson, N.M., Riby, P. & Lepp, N.W.: Arsenic mobility in brownfield soils amended with green waste compost or biochar and planted with Miscanthus. *Environ. Pollut.* 157 (2009b), pp. 2654–2662.

Hossain, M., Jahiruddin, M., Loeppert, R., Panaullah, G., Islam, M. & Duxbury, J.: The effects of iron plaque and phosphorus on yield and arsenic accumulation in rice. *Plant Soil* 317(2009), pp. 167–176.

Hossain, M.F.: Arsenic contamination in Bangladesh – An overview. *Agr. Ecosyst. Environ.* 113 (2006), pp. 1–16.

Huang, J.W., Poynton, C.Y., Kochian, L.V. & Elless, M.P.: Phytofiltration of arsenic from drinking water using arsenic-hyperaccumulating ferns. *Environ. Sci. Technol.* 38 (2004), pp. 3412–3417.

Jankong, P., Visoottiviseth, P. & Khokiattiwong, S.: Enhanced phytoremediation of arsenic contaminated land. *Chemosphere* 68 (2007), pp. 1906–1912.

Juwarkar, A.A., Yadav, S.K., Kumar, P. & Singh, S.K.: Effect of biosludge and biofertilizer amendment on growth of *Jatropha curcas* in heavy metal contaminated soils. *Environ. Monit. Assess.* 145 (2008), pp. 7–15.

Kertulis, G.M., Ma, L.Q., MacDonald, G.E., Chen, R., Winefordner, J.D. & Cai, Y.: Arsenic speciation and transport in *Pteris vittata* L. and the effects on phosphorus in the xylem sap. *Environ. Exp. Bot.* 54 (2005), pp. 239–247.

Kertulis-Tartar, G.M., Ma, L.Q., Tu, C. & Chirenje, T.: Phytoremediation of an arsenic-contaminated site using *Pteris vittata* L.: a two-year study. *Int. J. Phytorem.* 8 (2006), pp. 311–322.

Khan, M.A., Stroud, J.L., Zhu, Y.-G., McGrath, S.P. & Zhao, F.J.: Arsenic bioavailability to rice is elevated in Bangladeshi paddy soils. *Environ. Sci. Technol.* 44 (2010), pp. 8515–8521.

Kim, J.Y., Davis, A.P. & Kim, K.W.: Stabilization of available arsenic in highly contaminated mine tailings using iron. *Environ. Sci. Technol.* 37 (2003), pp. 189–195.

Kumpiene, J., Lagerkvist, A. & Maurice, C.: Stabilization of As, Cr, Cu, Pb and Zn in soil using amendments– A review. *Waste Manage.* 28 (2008), pp. 215–225.

Li, R.Y., Stroud, J.L., Ma, J.F., McGrath, S.P. & Zhao, F.J.: Mitigation of arsenic accumulation in rice with water management and silicon fertilization. *Environ. Sci. Technol.* 43 (2009), pp. 3778–3783.

Liu, W.J., Zhu, Y.G., Hu, Y., Williams, P.N., Gault, A.G., Meharg, A.A., Charnock, J.M. & Smith, F.A.: Arsenic sequestration in iron plaque, its accumulation and speciation in mature rice plants (*Oryza Sativa* L.). *Environ. Sci. Technol.* 40 (2006), pp. 5730–5736.

Luongo, T. & Ma, L.Q.: Characteristics of arsenic accumulation by Pteris and non-Pteris ferns. *Plant Soil* 277 (2005), pp. 117–126.

Ma, J.F. & Yamaji, N.: Silicon uptake and accumulation in higher plants. *Trends Plant Sci.* 11 (2006), pp. 392–397.

Ma, J.F., Yamaji, N., Mitani, N., Xu, X.Y., Su, Y.H., McGrath, S.P. & Zhao, F.J.: Transporters of arsenite in rice and their role in arsenic accumulation in rice grain. *P. Natl. Acad. Sci. USA* 105 (2008), pp. 9931–9935.

Ma, L.Q., Komar, K.M., Tu, C., Zhang, W. & Cai, Y.: A fern that hyperaccumulates arsenic. *Nature* 409 (2001), p. 579.

Mahmud, R., Inoue, N., Kasajima, S. & Shaheen, R.: Assessment of potential indigenous plant species for the phytoremediation of arsenic-contaminated areas of Bangladesh. *Int. J. Phytorem.* 10 (2008), pp. 119–132.

Mains, D., Craw, D., Rufaut, C.G. & Smith, C.M.S.: Phytostabilization of gold mine tailings from New Zealand. Part 2: Experimental evaluation of arsenic mobilization during revegetation. *Int. J. Phytorem.* 8 (2006), pp. 163–183.

Mathews, S., Ma, L.Q., Rathinasabapathi, B., Natarajan, S. & Saha, U.K.: Arsenic transformation in the growth media and biomass of hyperaccumulator *Pteris vittata* L. *Bioresour. Technol.* 101 (2010), pp. 8024–8030.

McGrath, S.P. & Zhao, F.J.: Phytoextraction of metals and metalloids from contaminated soils. *Curr. Opin. Biotechnol.* 14 (2003), pp. 277–282.

Mei, X.Q., Ye, Z.H. & Wong, M.H.: The relationship of root porosity and radial oxygen loss on arsenic tolerance and uptake in rice grains and straw. *Environ. Pollut.* 157 (2009), pp. 2550–2557.

Mohan, D., Pittman Jr, C.U., Bricka, M., Smith, F., Yancey, B., Mohammad, J., Steele, P.H., Alexandre-Franco, M.F., Gáez-Serrano, V. & Gong, H.: Sorption of arsenic, cadmium, and lead by chars produced from fast pyrolysis of wood and bark during bio-oil production. *J. Colloid Interface Sci.* 310 (2007), pp. 57–73.

Namgay, T., Singh, B. &Singh, B.P.: Plant availability of arsenic and cadmium as influenced by biochar application to soil. *19th World Congress of Soil Science, Soil Solutions for a Changing World*, 2010.

Natarajan, S., Stamps, R.H., Saha, U.K. & Ma, L.Q.: Effects of nitrogen and phosphorus levels, and frond-harvesting on absorption, translocation and accumulation of arsenic by Chinese Brake fern (*Pteris Vittata* L.). *Int. J. Phytoremediat.* 11 (2009), pp. 313–328.

Niazi, N.K., Singh, B., Zwieten, L.V. & Kachenko, A.G.: Arsenic hyperaccummulation by ferns: A field study in northern NSW. *19th World Congress of Soil Science, Soil Solutions for a Changing World*, 2010.

Poynton, C.Y., Huang, J.W., Blaylock, M.J., Kochian, L.V. & Elless, M.P.: Mechanisms of arsenic hyperaccumulation in Pteris species: root As influx and translocation. *Planta* 219 (2004), pp. 1080–1088.

Rahman, M. A., Hasegawa, H. & Ueda, K.: Arsenic accumulation in duckweed (*Spirodela polyrhiza* L.): A good option for phytoremediation. *Chemosphere* 69 (2007), pp. 493–499.

Salt, D.E, Smith, R.D., & Raskin, I.: Phytoremediation. *Annu. Rev. Plant Physiol. Mol. Biol.* 49 (1998), pp. 643–668.

Santos, J.A.G., Gonzaga, M.I.S., Ma, L.Q. & Srivastava, M.: Timing of phosphate application affects arsenic phytoextraction by *Pteris vittata* L. of different ages. *Environ. Pollut.* 154 (2008), pp. 306–311.

Sarkar, D., Datta, R. & Sharma, S.: Fate and bioavailability of arsenic in organo-arsenical pesticide-applied soils. Part-I: incubation study. *Chemosphere* 60 (2005), pp. 188–195.

Sas-Nowosielska, A., Kucharski, R., Malkowski, E., Pogrzeba, M., Kuperberg, J.M. & Krynski, K.: Phytoextraction crop disposal – an unsolved problem. *Environ. Pollut.* 128 (2004), pp. 373–379.

Sharples, J.M., Meharg, A.A., Chambers, S.M. & Cairney, J.W.G.: Symbiotic solution to arsenic contamination. *Nature* 404 (2000), pp. 951–952.

Shaw, J.K., Fathordoobadi, S., Zelinski, B.J., Ela, W.P. & Sáez, A.E.: Stabilization of arsenic-bearing solid residuals in polymeric matrices. *J. Hazard. Mater.* 152 (2008), pp. 1115–1121.

Shelmerdine, P.A., Black, C.R., McGrath, S.P. & Young, S.D.: Modelling phytoremediation by the hyper-accumulating fern, *Pteris vittata*, of soils historically contaminated with arsenic. *Environ. Pollut.* 157 (2009), pp. 1589–1596.

Shiralipour, A., Ma, L., & Cao, R.: Effects of compost on arsenic leachability in soils and arsenic uptake by a fern. Florida Centre for Solid Hazardous Waste Management, State University System of Florida, Gainesville, FL. Report, 2002, #02-04.

Srivastava M., Ma, L.Q., & Santos, J.A.G.: Three new arsenic hyperaccumulating ferns. *Sci. Total Environ.* 364 (2006), pp. 24–31.

Su, Y.H., McGrath, S.P., Zhu, Y.G. & Zhao, F.J.: Highly efficient xylem transport of arsenite in the arsenic hyperaccumulator *Pteris vittata*. *New Phytol.* 180 (2008), pp. 434–441.

Trotta, A., Falaschi, P., Cornara, L., Minganti, V., Fusconi, A., Drava, G. & Berta, G.: Arbuscular mycorrhizae increase the arsenic translocation factor in the As hyperaccumulating fern *Pteris vittata* L. *Chemosphere* 65 (2006), pp. 74–81.

Tu, C. & Ma, L.Q.: Effects of arsenate and phosphate on their accumulation by an arsenic-hyperaccumulator *Pteris vittata* L. *Plant Soil* 249 (2003), pp. 373–382.

Tu, C. Ma, L.Q., & Bondada, B.: Arsenic accumulation in the hyperaccumulator Chinese brake and its utilization potential for phytoremediation. *J. Environ. Qual.* 31 (2002), pp. 1671–1675.

Tu, C., Ma, L.Q., Zhang, W., Cai, Y. & Harris, W.G.: Arsenic species and leachability in the fronds of the hyperaccumulator Chinese brake (*Pteris vittata* L.). *Environ. Pollut.* 124 (2003), pp. 223–230.

Tu, S., Ma, L.Q., Fayiga, A.O. & Zillioux, E.J.: Phytoremediation of arsenic-contaminated groundwater by the arsenic hyperaccumulating fern *Pteris vittata* L. *Int. J. Phytoremediat.* 6 (2004a), pp. 35–47.

Tu, S., Ma, L.Q. & Luongo, T.: Root exudates and arsenic accumulation in arsenic hyperaccumulating *Pteris vittata* and non-hyperaccumulating *Nephrolepis exaltata*. *Plant Soil* 258 (2004b), pp. 9–19.

Tu, S., Ma, L.Q., MacDonald, G.E. & Bondada, B.: Effects of arsenic species and phosphorus on arsenic absorption, arsenate reduction and thiol formation in excised parts of *Pteris vittata* L. *Environ. Exp. Bot.* 51 (2004c), pp. 121–131.

Ultra, V.U., Nakayama, A., Tanaka, S., Kang, Y.M., Sakurai, K. & Iwasaki, K.: Potential for the alleviation of arsenic toxicity in paddy rice using amorphous iron-(hydr)oxide amendments. *Soil Sci. Plant Nutr.* 55 (2009), pp. 160–169.

Vamerali, T., Bandiera, M., Coletto, L., Zanetti, F., Dickinson, N.M. & Mosca, G.: Phytoremediation trials on metal- and arsenic-contaminated pyrite wastes (Torviscosa, Italy). *Environ. Pollut.* 157 (2009), pp. 887–894.

Vangronsveld, J., Herzig, R., Weyens, N., Boulet, J., Adriaensen, K., Ruttens, A., Thewys, T., Vassilev, A., Meers, E., Nehnevajova, E., van der Lelie, D. & Mench, M.: Phytoremediation of contaminated soils and groundwater: lessons from the field. *Environ. Sci. Pollut. Res.* 16 (2009), pp. 765–794.

Vazquez, S., Agha, R., Granado, A., Sarro, M.J., Esteban, E., Penalosa, J.M. & Carpena, R.O.: Use of white lupin plant for phytostabilization of Cd and As polluted acid soil. *Water Air Soil Poll.* 177 (2006), pp. 349–365.

Visoottiviseth, P., Francesconi, K & Sridokchan, W.: The potential of Thai indigenous plant species for the phytoremediation of arsenic contaminated land. *Environ. Pollut.* 118 (2002), pp. 453–461.

Wang, H.B., Wong, M.H., Lan, C.Y., Baker, A.J.M., Qin, Y.R., Shu, W.S., Chen, G.Z. & Ye, Z.H.: Uptake and accumulation of arsenic by 11 Pteris taxa from southern China. *Environ. Pollut.* 145 (2007), pp. 225–233.

Wang, J.R., Zhao, F.J., Meharg, A.A., Raab, A., Feldmann, J. & McGrath, S.P.: Mechanisms of arsenic hyperaccumulation in *Pteris vittata*. Uptake kinetics, Interactions with phosphate, and Arsenic speciation. *Plant Physiol.* 130 (2002), pp. 1552–1561.

Warren, G.P., Alloway, B.J., Lepp, N.W., Singh, B., Bochereau, F.J.M. & Penny, C.: Field trials to assess the uptake of arsenic by vegetables from contaminated soils and soil remediation with iron oxides. *Sci. Total Environ.* 311 (2003), pp. 19–33.

Whiting, S.N., Reeves, R.D., Richards, D., Johnson, M.S., Cooke, J.A., Malaisse, F., Paton, A., Smith, J.A.C., Angle, J.S., Chaney, R.L., Ginocchio, R., Jaffr, T., Johns, R., McIntyre, T., Purvis, O.W., Salt, D.E., Schat, H., Zhao, F.J. & Baker, A.J.M.: Research priorities for conservation of metallophyte biodiversity and their potential for restoration and site remediation. *Restor. Ecol.* 12 (2004), pp. 106–116.

Williams, P.N., Villada, A., Deacon, C., Raab, A., Figuerola, J., Green, A.J., Feldmann, J. & Meharg, A.A.: Greatly enhanced arsenic shoot assimilation in rice leads to elevated grain levels compared to wheat and barley. *Environ. Sci. Technol.* 41 (2007), pp. 6854–6859.

Williams, P.N., Lei, M., Sun, G., Huang, Q., Lu, Y., Deacon, C., Meharg, A.A. & Zhu, Y.G.: Occurrence and partitioning of cadmium, arsenic and lead in mine impacted paddy rice: Hunan, China. *Environ. Sci. Technol.* 43 (2009), pp. 637–642.

Wu, F., Ye, Z., Wu, S. & Wong, M.: Metal accumulation and arbuscular mycorrhizal status in metallicolous and nonmetallicolous populations of *Pteris vittata* L. and *Sedum alfredii Hance. Planta* 226 (2007), pp. 1363–1378.

Wu, F.Y., Ye, Z.H. & Wong, M.H.: Intraspecific differences of arbuscular mycorrhizal fungi in their impacts on arsenic accumulation by *Pteris vittata* L. *Chemosphere* 76 (2009), pp. 1258–1264.

Xu, X.Y., McGrath, S.P., Meharg, A.A. & Zhao, F.J.: Growing rice aerobically markedly decreases arsenic accumulation. *Environ. Sci. Technol.* 42 (2008), pp. 5574–5579.

Yan, X.L., Chen, T.B., Liao, X.Y., Huang, Z.C., Pan, J.R., Hu, T.D., Nie, C.J. & Xie, H.: Arsenic transformation and volatilization during incineration of the hyperaccumulator *Pteris vittata* L. *Environ. Sci. Technol.* 42 (2008), pp. 1479–1484.

Zhao, F.J. & McGrath, S.P.: Biofortification and phytoremediation. *Curr. Opin. Plant Biol.* 12 (2009), 373–380.

Zhao, F.J., Dunham, S.J. & McGrath, S.P.: Arsenic hyperaccumulation by different fern species. *New Phytol.* 156 (2002), pp. 27–31.

Zhao, F.J., Ma, J.F., Meharg, A.A. & McGrath, S.P.: Arsenic uptake and metabolism in plants. *New Phytol.* 181 (2009), pp. 777–794.

Zhao, F.J., McGrath, S.P. & Meharg, A.A.: Arsenic as a food chain contaminant: mechanisms of plant uptake and metabolism and mitigation strategies. *Annu. Rev. Plant Biol.* 61 (2010), pp. 535–559.

CHAPTER 5

Fundamentals of electrokinetics

Soon-Oh Kim, Keun-Young Lee & Kyoung-Woong Kim

5.1 INTRODUCTION

Numerous sites are contaminated due to various activities such as improper treatment of wastes, abandoned mining wastes, leakage of landfill leachate, military activities, impertinent collection of used batteries, and accidental spills. Contamination frequently affects large areas of soil underlying the surface, and contaminants found in these areas include a wide range of toxic chemicals such as heavy metals, metalloids, radionuclides, and organic compounds. In addition, there are various types of contaminated lands such as paddy soils, farmlands, factory sites, mine fields, and residential districts. Toxic contaminants within and migrating from these lands threaten human health through their detrimental effects on agricultural crops and drinking water reservoirs, including groundwater and surface water in the local area. Accordingly, the remediation of contaminated sites has become a great environmental concern and urgent priority. Development of high-performance technologies for cleaning up these sites is one of the most important technological challenges.

Until recently, a variety of technologies have been developed to remediate soils, sediments, and groundwater, and they can generally be classified into three groups based on the principles applied. The first one is biological remediation, which has been mainly used to detoxify contaminants, particularly organic toxicants. The technology includes phytoremediation, bioventing/biosparging, biopile, landfarming, composting, biofilter, and *in-situ* bioremediation. The second group is physico-chemical decontamination, which has been usually applied to remove inorganic contaminants, especially heavy metals. This includes excavation, soil washing/flushing, solidification and stabilization (SS), soil vapor extraction (SVE), and electrokinetics. The third decontamination method is based on thermal principles and includes incineration, thermal desorption, vitrification, and thermally enhanced SVE.

Even though a number of remediation technologies have been developed so far, most of them are costly, energy intensive, ineffective, and could cause adverse and secondary environmental impacts. A major limitation to most successful remediation technologies, such as SVE and soil washing/flushing, is that they are restricted to soils with high hydraulic conductivity. In other words, they cannot be effectively used to clean up low-permeability and fine-grained soils and sediments. Furthermore, most of the technologies have been demonstrated to be inadequate for remediation of soils affected by a mixture of contaminants, such as heavy metals coexisting with organic contaminants. In contrast, electrokinetics has been recognized as one of the most promising technologies to overcome such limitations and difficulties.

Electrokinetics is also termed electrokinetic soil processing, electrokinetic remediation, electroreclamation, electrochemical remediation, and electrochemical decontamination. The electrokinetics needs low-level current, which is normally direct electricity, except for some special applications, in order of $mA\,cm^{-2}$ of a cross-sectional area between the electrodes to remove contaminants from soils. The low-level direct current results in physico-chemical and hydrological changes in the soil mass, leading to species transport by coupled mechanisms.

The electrokinetics is an effective technology for removing contaminants in low-permeability soils ranging from clay to clayey sand. The advantages of this technology include its low cost of operation and its potential applicability to a wide range of contaminants (Alshawabkeh *et al.*, 1999; Kim and Kim, 2002; Pamukcu and Wittle, 1994; Reddy and Cameselle, 2009a). The electrokinetic technology is envisioned for the removal/separation of organic and inorganic contaminants and radionuclides. The potential of the technology for soil remediation has resulted in several field implementations (DOE, 1998a; 1998b; Gent *et al.*, 2004; Ho *et al.*, 1999a; 1999b; Lageman, 1993; USEPA, 1998; 2003). Except for the applicability of electrokinetics to soil remediation, the technology was originally used to consolidate low-permeability clayey soils for a long time (Mitchell, 1991; 1993; Mitchell and Soga, 2005). In this chapter, the application of electrokinetics for soil remediation is discussed.

The purpose of this chapter is to provide the fundamental concept of electrokinetics in soil remediation. First, a variety of electrokinetic phenomena that occur in soils when an electric field is applied are introduced. These are important because the electrokinetic technology is based on naturally-occurring phenomena. Next, the fundamental principles and implementation system of electrokinetics are detailed as well. In addition, the unique advantages and disadvantages of electrokinetics over other remediation technologies are addressed. In the second part of the chapter, the main considerations and operational parameters affecting the performance of electrokinetic processing are detailed. In the third part, the researches and projects on electrokinetics that have been conducted are presented and summarized with a focus on field applications. Finally, the chapter closes with a brief consideration of prospects to improve the performance of electrokinetics in soil remediation.

5.2   ELECTROKINETIC PHENOMENA

When electricity is supplied to a wet porous medium, such as soil, a variety of physico-chemical phenomena occur, and they are called electrokinetic phenomena. In the early 19th century, these electrokinetic phenomena were identified by Russian scientist Reuss (1809), who conducted an experiment in which direct electricity was applied to wet clay soil (Acar and Alshawabkeh, 1993). However, it was in the early 20th century that the scientific phenomena became a basis for engineering technologies. The electrokinetic phenomena accompany various physical and chemical interactions and other related phenomena, and it is impossible to address all of them in details here. Please refer to Mitchell (1991; 1993), Probstein (2003), and Mitchell and Soga (2005) for an in-depth study.

In this section, representative electrokinetic phenomena that are directly related to the principles of electrokinetic technology as well as the mechanisms of contaminant transport are discussed. Electrolytic reaction of water is also addressed, because the reaction greatly affects the interactions between soil particles and contaminants and finally controls transport of contaminants in soil media. The electrokinetic phenomena related to contaminant transport are electromigration (also termed ionic migration), electroosmosis (electroosmotic advection), electrophoresis, and diffusion, and these are bases of electrokinetic technology by which contaminants are removed from soils.

5.2.1   *Electrokinetic transport phenomena*

5.2.1.1   *Electromigration or ionic migration*
Under the influence of a direct current (DC) electric field, charged chemical species, such as ions and polar molecules, move in pore water within soil media, and this transport phenomenon is termed electromigration or ionic migration. On the basis of electromigration, cationic and anionic contaminants are removed in the cathode (negative electrode) and anode (positive electrode), respectively, during electromigration. Ionic mobility u_i [$m^2 \, s^{-1} \, V^{-1}$], defines the transport velocity of the ionic species i under the effect of a unit electric field and can be theoretically estimated by using the Nernst-Townsend-Einstein relation (Holmes, 1962) between the diffusion

Table 5.1. Diffusion coefficient, ionic mobility at infinite dilution and effective ionic mobility in soil.

Metal species	Molecular diffusion coefficient $D_i \times 10^{10}$ $[m^2 s^{-1}]^{1)}$	Ionic mobility $u_i \times 10^8$ $[m^2 s^{-1} V^{-1}]$	Effective ionic mobility $u_i^* \times 10^9$ $[m^2 s^{-1} V^{-1}]^{2)}$
Li^+	10.3	4.01	5.62
Na^+	13.3	5.18	7.25
K^+	19.6	7.63	10.7
Rb^+	20.7	8.06	11.3
Cs^+	20.5	7.98	11.2
Be^{2+}	5.98	4.66	6.52
Mg^{2+}	7.05	5.49	7.69
Ca^{2+}	7.92	6.17	8.64
Sr^{2+}	7.9	6.15	8.61
Ba^{2+}	8.46	6.59	9.22
Pb^{2+}	9.25	7.20	10.09
Cu^{2+}	7.13	5.55	7.77
Fe^{2+}	7.19	5.60	7.84
Cd^{2+}	7.17	5.58	7.82
Zn^{2+}	7.02	5.47	7.65
Ni^{2+}	6.79	5.29	7.40
Fe^{3+}	6.07	7.09	9.93
Cr^{3+}	5.94	6.94	9.72
Al^{3+}	5.95	6.95	9.73

[1]Values from Mitchell (1993) and Mitchell and Soga (2005).
[2]Values calculated using 0.4 porosity (n) and 0.35 tortuosity (τ) (Kim et al., 2009a).

coefficient, D_i $[m^2 s^{-1}]$, in free solution and ionic mobility, u_i (Alshawabkeh and Acar, 1996; Kim et al., 2002a; Mitchell and Soga, 2005):

$$u_i = (D_i |z_i| F)/(RT) \tag{5.1}$$

where z_i is the charge of species i, F is Faraday's constant (96,485 C/mole electrons), R is the universal gas constant (8.3144 J K^{-1} mole^{-1}), and T is the absolute temperature [K]. The ionic mobilities in the free solution range from 10^{-8} to $10^{-7} m^2 s^{-1} V^{-1}$, except for protons ($4 \times 10^{-7} m^2 s^{-1} V^{-1}$) and hydroxyl ions ($2 \times 10^{-7} m^2 s^{-1} V^{-1}$). However, the effective ionic mobility, u_i^* $[m^2 s^{-1} V^{-1}]$ in soil is substantially lower relative to that (u_i) in solution due to the effect of porosity (n) and tortuosity (τ). The effective mobility, u_i^*, is defined as follows:

$$u_i^* = n\tau u_i \tag{5.2}$$

where n and τ are the porosity and tortuosity of soil, respectively. The tortuosity factors reported in different studies are as low as 0.01 and as high as 0.84, and they usually range from 0.20 to 0.50 (Shackelford, 1990). Table 5.1 presents the diffusion coefficients, ionic mobilities, and effective ionic mobilities of representative metal ions. The typical effective ionic mobilities in clayey soils, which are targeted to electrokinetics, range from 3×10^{-9} to $1 \times 10^{-8} m^2 s^{-1} V^{-1}$ (Mitchell, 1991; 1993; Mitchell and Soga, 2005). The effective ionic mobility is affected by several factors, such as the electric conductivity and ionic strength of the pore water, the charge of the ion, and temperature. For electromigrative velocity of species i in soils, $V_{em,i}$, can be calculated by using effective ionic mobility, u_i^*, and electric field strength E $[V m^{-1}]$:

$$V_{em,i} = u_i^* E \tag{5.3}$$

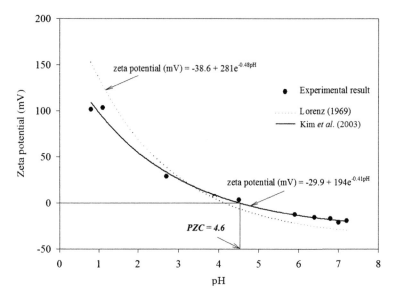

Figure 5.1.   Empirical relation between pH and zeta potential (ζ).

5.2.1.2   Electroosmosis or electroosmotic advection

During electrokinetic processing, pore water is transported by electroosmosis, which together with electromigration, provides an important mechanism for removing contaminants from porous media. Electroosmotic velocity on a plane surface, V_{eo} [m s^{-1}], is expressed as (Mitchell and Yeung, 1991; Mitchell and Soga, 2005; Pamucku and Wittle, 1994; Probstein, 2003; Shapiro, 1990):

$$V_{eo} = -(\varepsilon\zeta/\mu)E \qquad (5.4)$$

where ε is the permittivity of the medium [C V^{-1} m^{-1}], ζ is the zeta potential [V], E is the electric field strength or electrical potential gradient [V m^{-1}] in a direction parallel to the electroosmotic flow, and μ is the viscosity of the medium [N s m^{-2}]. The zeta potential is defined as the electrical potential at the shear or slipping plane between the moving and stationary phases (Dzombak and Morel, 1990; Mitchell and Soga, 2005). This formula for electroosmotic velocity on a charged plane surface is known as the Helmholtz-Smoluchowski relation. According to this equation, the electroosmotic velocity (V_{eo}) is significantly affected by the electric field strength (E) and the zeta potential (ζ). The magnitude and sign of ζ depend on the interfacial chemistry, and it can be expressed by a complex function of the chemistry of both liquid and solid phases (Eykholt and Daniel, 1994). An example of the relationship between the zeta potential and pH of a kaolinite medium has been empirically determined (Eykholt and Daniel, 1994; Kim et al., 2003; Lorenz, 1969):

$$\zeta \, [\text{mV}] = -38.6 + 281e^{-0.48\text{pH}} \qquad (5.5a)$$

$$\zeta \, [\text{mV}] = -29.9 + 194e^{-0.41\text{pH}} \qquad (5.5b)$$

The empirical relations between pH and ζ are presented in Figure 5.1, which are compared to those suggested by Lorenz (1969) and Kim et al. (2003). In addition, the effect of electrolyte concentration on zeta potential is represented by Kruyt (1952) and Alshawabkeh (1994):

$$\zeta = A - B\log C \qquad (5.6)$$

Table 5.2. Electroosmotic permeability coefficient[1].

Material	Water content [%]	$k_{eo} \times 10^{-9}$ [m^2 s^{-1} V^{-1}]	Approximate k_h [m s^{-1}][2]
London clay	52.3	5.8	10^{-10}
Boston blue clay	50.8	5.1	10^{-10}
Kaolin	67.7	5.7	10^{-9}
Clayey silt	31.7	5.0	10^{-8}
Na-Montmorillonite	170	2.0	10^{-11}
	2000	12.0	10^{-10}
Fine sand	26.0	4.1	10^{-6}
Silty clay, West Branch Dam	32.0	3.0–6.0	10^{-10}
Clayey silt, Little Pic River, Ontario	26.0	1.5	10^{-7}

[1]From Mitchell and Soga (2005).
[2]Hydraulic conductivity.

where A and B are experimentally determined constants, and C [mole m^{-3}] is the total electrolyte concentration. Accordingly, the zeta potential is controlled by the pH of the medium and the total electrolyte concentration in the pore water. The point where the plot of the zeta potential *versus* pH passes through the zero zeta potential is called the isoelectric point or Point of Zero Charge (PZC).

The electroosmotic flow rate, Q [m^3 s^{-1}], is expressed as:

$$Q = nV_{eo}A \tag{5.7}$$

where n and A [m^2] are the porosity and the cross-sectional area of the soil. From Equations (5.4) and (5.7), we obtain:

$$Q = (-(n\varepsilon\zeta)/\mu)EA \tag{5.8}$$

Similar to Darcy's law, Equation (5.8) can be expressed as:

$$Q = -k_{eo}EA \tag{5.9}$$

where the coefficient k_{eo} [m^2 s^{-1} V^{-1}], is termed electroosmotic permeability. Comparing Equation (5.8) with Equation (5.9), the electroosmotic permeability coefficient (k_{eo}) is given as:

$$k_{eo} = (n\varepsilon\zeta)/\mu \tag{5.10}$$

Mitchell and Soga (2005) reported that k_{eo} ranges from 1×10^{-9} to 1×10^{-8} m^2 s^{-1} V^{-1} (Table 5.2). According to the Helmholtz-Smoluchowski theory and Equation (5.10), k_{eo} should be relatively independent of pore size, and this is shown in Table 5.2, but the hydraulic conductivity (k_h) varies as the square of some effective pore size (Mitchell and Soga, 2005). Electroosmosis acts more effectively in flowing water through fine-grained soil than hydraulically-driven flow, since k_{eo} is independent of pore size. It is easily demonstrated by the following example proposed by Mitchell and Soga (2005). One can consider a fine sand and a clay of k_h of 1×10^{-5} m s^{-1} and 1×10^{-10} m s^{-1}, respectively, and it is assumed that both soils have k_{eo} values of 5×10^{-9} m^2 s^{-1} V^{-1}. If hydraulic flow rates are equal in both soils, then $(i_h \times k_h) = (E \times k_{eo})$, where E is the electrical potential gradient or electric field strength. Hence, $i_h = (k_{eo}/k_h)E$. If E of 20 V m^{-1} is applied, i_h is 0.01 for the fine sand and 1000 for the clay. It means that a hydraulic gradient of only 0.01 can move water as effectively as an electrical gradient of 20 V m^{-1} in fine sand. However, for the clay, a hydraulic gradient of 1000 would be needed to offset the electroosmotic flow (Mitchell and Soga, 2005).

As shown in Equation (5.8), there are a number of factors affecting the electroosmotic flow rate. Even though the permittivity (ε) and the viscosity (μ) of pore water are changed by the concentration of dissolved constituents and temperature, they are generally considered as constant values (Eykholt and Daniel, 1994; Kim, 2001; Shapiro, 1990). As discussed briefly above, however, the magnitude and sign of the ζ significantly depends on several factors such as pH and the ionic strength of the pore water, speciation of chemical constituents, temperature, and type of soil (Page and Page, 2002; Virkutyte et al., 2002). In particular, the pH and ionic strength of the pore water are crucial factors impacting ζ (Alshawabkeh et al., 1999). Since soil surface is negatively charged in common conditions, the ζ of soil surface appears to be negative. Accordingly, the electroosmosis takes place from anode towards cathode. However, the magnitude and sign of ζ continuously changes during the duration of the electrokinetic process because the pH and ionic strength of the pore water are not constant (Shapiro, 1990). Therefore, the direction and flow rate of electroomosis do not remain constant. The negative value of zeta potential gradually increases with the decrease in soil pH and finally becomes positive (Fig. 5.1). If the sign of ζ changes, then the direction of electroosmosis is reversed (Kim et al., 2002a). When electroosmosis occurs from anode towards cathode, generally speaking, it is called normal electroosmosis. On the contrary, the direction of reversal electroosmosis is from cathode to anode (Kim, 2001; Probstein, 2003; Reddy and Camesselle, 2009a).

5.2.1.3   Electrophoresis

When a direct current electric field is imposed on charged clay particles or colloids, those charged particles and bound contaminants are electrostatically attracted to one of the electrodes and are repelled from the other (Mitchell and Soga, 2005; Probstein, 2003). For example, negatively charged clay particles move towards the anode. This movement of charged solid particles or colloids is called electrophoresis or cataphoresis (Reddy and Cameselle, 2009a). Compared with electromigration and electroosmosis, electrophoretic transport is not significant and is negligible in a compact system such as low-permeability soil, because colloids and solid particles are large in size and exhibit higher frictional drag force in motion. Meanwhile, electrophoretic transport becomes crucial and dominant for biocolloids (e.g., bacteria) and micelles in soil suspension systems (Reddy and Cameselle, 2009a).

5.2.1.4   Diffusion

If a concentration gradient exists in a soil system, ionic and molecular contaminants transport from areas of higher concentration to areas of lower concentration. The transport phenomenon is called diffusion (Mitchell and Soga, 2005; Probstein, 2003). As shown in Equation (5.1) and Table 5.1, the ionic mobility (u_i) of a charged species i is much greater than its diffusion coefficient (D_i), and contaminant transport by diffusion is neglected in the electrokinetic process.

5.2.2   Electrolysis of water

When inert electrodes are placed in water and a direct current is passed through them, water molecules are electrolyzed at the surface of electrodes. As a result, hydrogen ions are produced at the anode and hydroxide ions at the cathode. At the electrodes, electrolysis of water takes place as follows (Kim and Kim, 2002; Reddy and Cameselle, 2009a):

$$\text{At anode: } 2H_2O - 4e^- \rightarrow 4H^+ + O_2(g) \tag{5.11}$$

$$\text{At cathode: } 2H_2O + 2e^- \rightarrow 2OH^- + H_2(g) \tag{5.12}$$

The production of H^+ ions at the anode decreases the pH by Equation (5.11), but the reaction of Equation (5.12) increases the pH at the cathode by increasing OH^- ions. This electrolysis results in an acid front at the anode and an alkaline front at the cathode. These fronts travel through soil media and finally move towards the cathode and the anode, respectively. The propagation of the acid and the base front through soil promotes the dissolution of metal ions near the anode and the precipitation of the metal ions near the cathode (Alshawabkeh, 1994; Kim *et al.*, 2002a). These conditions significantly affect the pH and ionic strength of pore water, the mobility and solubility of metal contaminants, and charge conditions of soil particles. Comparing Equations (5.11) and (5.12), the concentration of H^+ ions is two times higher than that of OH^- ions because the electrolysis of 1 mole of water molecule renders 2 moles of H^+ions and 1 mole of OH^- ions. In addition, the ionic mobility of the hydrogen ions is twice as high as that of the hydroxide ion, as addressed in Section 5.2.1.1. Since normal electroosmosis (i.e., from anode toward cathode) occurs in common pH conditions of soils, the electromigration of hydrogen ions is enhanced by electroosmosis toward the cathode. For these reasons, the acid front is dominant in soils and decreases soil pH. A decrease in soil pH is favorable in the electrokinetic removal of cationic contaminants, such as toxic heavy metals, because of effective desorption and dissolution of contaminants that are adsorbed on the soil surface and precipitated within soil. Having greater replacing power on the soil surface than any other species, hydrogen ions in the acid front displace metal ions and organics adsorbed on the soil surface, resulting in more mobile forms of contaminants in pore water that can be transported by electromigration. The variation of pH conditions in soils by electrolysis of water in the electrode compartment has effects on ionic strength of pore water and soil surface properties such as cation or anion exchange capacity, magnitude and sign of the electrokinetic zeta potential. Furthermore, speciation, mobility and solubility of contaminants are often varied with pH in soils during treatment, which may limit or enhance treatment efficiencies (Kim *et al.*, 2002a).

5.2.3 *Fundamental principle of electrokinetic remediation*

5.2.3.1 *Transport and removal of inorganic contaminants*
The electrokinetic technology can be applied to remove inorganic contaminants from soils. Depending on the chemical and geochemical characteristics, inorganic contaminants are categorized into three groups: (i) cationic toxic heavy metals such as Cd, Cu, Hg, Ni, Pb, and Zn, (ii) anionic metals, metalloids, and other inorganics such as As, Cr, Se, NO_3, and F, (iii) radionuclides such as Sr and U (Kim and Kim, 2002; Reddy and Cameselle, 2009a). Their behaviors in soils widely vary depending on the type of contaminants and soil properties. Particularly, the speciation and transport of the inorganic contaminants are totally controlled by the dynamic changes in the pH and redox potential of the soil media under applied electric field during the electrokinetic processing (Virkutyte *et al.*, 2002). For example, as the acid front migrates through the soil bed, the cationic contaminants adsorbed on the soil surface are desorbed (Page and Page, 2002). The free chemical species present in the pore fluid transport towards the electrodes depending on their charge. Even though there has been some argument about the primary mechanism for removal of ionic contaminants, electromigration is considered as the dominant driving mechanism contributing to the transport of inorganic contaminants through the soil mass. However, both precipitation and sorption retard movement of cationic contaminants at high pH zones. Soil pH changes induced by the electric field complicate the geochemistry of soil and inorganic pollutant removal. Cations are collected at the cathode and anions at the anode as a result of the transport of chemical species in the soil pore fluid. In particular, metal and other cationic contaminants are removed from the soil with the cathode effluent solution and/or are deposited at the cathode. Hence, the effluent should be treated for the complete removal and the recovery of the metal contaminants.

5.2.3.2 *Transport and removal of organic contaminants*
Similar to inorganic contaminants, volatile and soluble organic contaminants can be effectively removed by electrokinetic technology. Recently, efforts have been focused on the electrokinetic removal of hydrophobic and persistent toxic organic compounds, such as polycyclic aromatic hydrocarbons (PAHs), polychlorinated organic compounds, pesticides, herbicides, and energetic compounds in soils (Reddy and Cameselle, 2009a). In the case of hydrophobic organic compounds (HOCs) with low solubility in water and with a high tendency to be adsorbed onto soil,

for example, electrokinetic remediation was previously considered as "not applicable" because transport by electroosmosis is not to be expected (Yang and Lee, 2009). However, the electrokinetic technology coupled with methods to increase the solubility of HOCs, has been developed and frequently applied. The PAHs, representative of HOCs, are very hydrophobic and have quite low aqueous solubility. The solubilization/desorption and fractionizing of PAHs in soil-water systems have been extensively studied using solubility-enhancing solutions such as surfactants and cosolvents to achieve effective removal of PAHs from contaminated sites (Yang and Lee, 2009). Electromigration is the dominant transport mechanism for the removal of inorganic contaminants, but electroosmosis becomes significant in removing organic contaminants due to their nonpolar characteristic. Refer to Reddy and Cameselle (2009b) for more detailed information about electrokinetic removal of organic compounds.

5.2.3.3   *Enhancement schemes for electrokinetic soil remediation*
In order to mobilize and solubilize contaminants, various enhancement techniques have been proposed and used (Alshawabkeh *et al.*, 1999; Kim and Kim, 2002; Page and Page, 2002). For inorganic contaminants, these include: (i) injection of enhancing agents such as acetic acid or use of a hydroxyl ion selective membrane in the cathode reservoir to prevent precipitation or to solubilize precipitates of cationic metal contaminants near the cathode, (ii) conditioning the anode and/or the cathode reservoirs to control pH and zeta potential, to enhance desorption, to increase the electroosmotic flow rate, and finally to increase mobility of contaminants, and (iii) adding or mixing strongly complexing agents, such as ammonia, citrate, and EDTA into soil, which compete with soil particles for metal contaminants to form soluble complexes. Among these enhancement technologies, the scheme for prevention of metal precipitation has been mostly focused and experimentally evaluated. Alshawabkeh (1994) summarized the characteristics of enhancement schemes as follows: (i) the precipitate should be solubilized and/or precipitation should be avoided; (ii) ionic conductivity across the soil specimen should not increase excessively in a short period of time both to avoid a premature decrease in electroosmotic transport and to allow transference of species of interest; (iii) the cathode reaction should possibly be depolarized to avoid generation of the hydroxide and its transport into the soil specimen; (iv) such depolarization will also assist in decreasing the electrical potential difference across the electrodes leading to lower energy consumption; (v) if any chemical is used, the precipitate of the metal with this new chemical should be perfectly soluble within the pH ranges attained, and (vi) any special chemicals introduced should not result in any increase in toxic residues in the soil mass. Recently, Yeung and Gu (2011) gave a comprehensive review on the techniques to enhance electrokinetic remediation of contaminated fine-grained soils. They summarized enhancement agents developed so far, such as chelants, complexing agents, surfactants and cosolvents, oxidizing/reducing agents, and cation solutions, and also explained the methods for electrode conditioning and use of an ion exchange membrane to control soil pH during the process.

5.2.3.4   *Implementation of electrokinetic remediation*
A typical system for electrokinetic remediation is presented in Figure 5.2. The system mainly consists of two unit processes: electrokinetic removal and effluent treatment processes. The implementation of electrokinetic removal of metal contaminants is exemplified here. Most of the metal contaminants are positive ionic compounds in a soil-water-electrolyte system, and they migrate towards the cathode when an electric field is applied to the system. Furthermore, these metal contaminants are removed through the cathode effluent solutions from the contaminated soils. The cathode effluent solutions should be properly treated in order to recover metals contained in these solutions. The typical ranges of electric field strength and current density for electrokinetic remediation are known to be $1–100\,V\,m^{-1}$ and $1–10\,A\,m^{-2}$, respectively (Alshawabkeh *et al.*, 1999). Although implementation of the electrokinetic remediation system is relatively simple, its design and operation for successful remediation is cumbersome due to complex dynamic electrochemical transport, transfer, and transformation processes that occur under applied electric potential. In

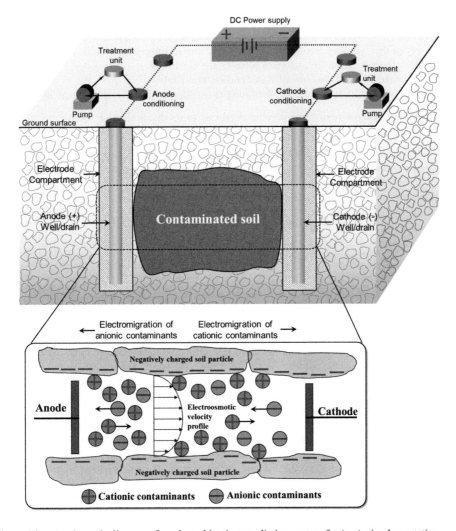

Figure 5.2.   A schematic diagram of an electrokinetic remediation system for *in-situ* implementation.

particular, the efficacy of electrokinetic remediation depends strongly on characteristics of contaminated media such as buffer capacity, mineralogy, and organic matter content, among others (Reddy and Cameselle, 2009a).

5.2.3.5   *Advantages and disadvantages of electrokinetic technology*
Electrokinetic remediation has been considered as a promising technology because of its unique advantages over other conventional technologies (Oonnittan *et al.*, 2009; Reddy and Cameselle, 2009a). These advantages include:

- Flexibility to use as *ex-situ* or *in-situ* method.
- Applicability to low-permeability and heterogeneous soils.

- Applicability to saturated and unsaturated soils.
- Applicability for heavy metals, metalloids, radionuclides, and organic contaminants, as well as in any of their combinations (contaminant mixtures or multiple contaminants).
- Feasibility to treat the entire soil mass between the electrodes.
- Effective controlling of the flow of water and contaminants.
- Easy integration with conventional technologies, including barrier and treatment systems.

Even though electrokinetics has been applied due to its discriminative merits, its ability to achieve remediation goals has yet to be addressed due to major disadvantages such as:

- Solubilization of major elements in soils and dissolution of mineral constituents of soils.
- Decrease in the efficiencies of electric current as well as contaminant removal due to non-target chemicals coexisting in soils.
- Post-treatment for land reclamation due to acidification and/or alkalification.
- Precipitation of metal species near the cathode.
- Problems due to electrode corrosion and excess soil heating.

5.3   DESIGN AND OPERATION OF ELECTROKINETIC REMEDIATION

There are a variety of factors and parameters affecting the performance of electrokinetic remediation (Alshawabkeh *et al.*, 1999; Page and Page, 2002; Virkutyte *et al.*, 2002). First, the types and features of soil and contaminants are crucial. The properties of soil evaluated prior to application of electrokinetic remediation include soil texture, porosity, electric conductivity, pH and pH buffering capacity, zeta potential of the surface, water content (degree of saturation), adsorption capacity, organic content, and mineralogy. The characteristics of contaminants targeted are type, concentration (initial level of contamination), ionic mobility, and chemical forms in soil (Alshawabkeh *et al.*, 1999). In addition, electrical parameters are important because they affect electromigration, electroosmosis, and electrolysis of water, and finally control the transport of contaminants. Hence, the overall efficiency and economy of the process are significantly influenced by electrical parameters. The electrical parameters, which are determined before application of electrokinetics, are electric field strength (electrical potential gradient) and current level (current density) (Kim and Kim, 2002). In order to improve the performance of electrokinetic remediation, the design of an electrokinetic system should be optimized prior to implementation. The representative design factors are related to the electrode, electrolyte, enhancement scheme, and type of electricity. The design of electrodes includes the distance between electrodes with different polarities (anode-cathode), the spacing of electrodes with equivalent polarity (anode-anode and cathode-cathode), the configuration (array) of electrodes, and the material and shape of electrodes (Alshawabkeh *et al.*, 1999). The pH and composition of electrolyte should be taken into account. The types of electricity are also important to improve the electrokinetic process. So far, the direct current (DC) has been usually used in electrokinetic remediation. However, the alternating current (AC) can be considered depending on the goal of the process (e.g., removal of volatile organic contaminants through soil heating). Even though the DC electric wave has been frequently used, in addition, one must evaluate which type of DC electric wave is most suitable to achieve the most efficient process because a variety of types of DC electric wave (e.g., full-wave, half-wave, pulse-wave with high frequency) can be generated by using different types of rectifier. In the next section, the factors affecting the performance of electrokinetic remediation are briefly discussed along with the parameters considered to optimize the operation and design of the process.

5.3.1   *Factors affecting the performance of electrokinetic remediation*

5.3.1.1   *Properties of soil*
A variety of soil properties affect electrokinetic performance, and they are divided into physical, chemical, and mineralogical features. The physical properties include texture or particle size

distribution (relative proportion of particle size), porosity, tortuosity, water content (or degree of saturation), activity, heterogeneity, and electrical properties, such as zeta potential of the surface, electroosmotic permeability and electric conductivity. In terms of chemical properties, adsorption capacity (or ion exchange capacity), pH, pH buffering capacity, and organic content should be evaluated prior to application of electrokinetic remediation. In addition, the mineralogy of the soil definitely affects the process. Particularly, the type and content of oxides, carbonates, and clay minerals contained in the soil is crucial because it affects numerous properties of soil such as adsorption capacity, electroosmotic permeability, and pH buffering capacity. All those properties of soils control electromigration of contaminants as well as electroosmosis, and finally affect the overall performance of the electrokinetic process. For example, soils of high water content, high degree of saturation, and low activity provide the most favorable conditions for transport of contaminants by electroosmotic advection and electromigration. However, soils of high activity (e.g., montmorillonite) exhibit high pH buffering capacity, and require excessive acid and/or enhancement agents to desorb and solubilize contaminants sorbed on the soil particle surface before they can be transported through the surface and removed (Alshawabkeh *et al.*, 1999). Consequently, the properties of soil are most crucial in electrokinetic remediation and should be evaluated prior to implementation.

5.3.1.2 *Characteristics of contaminants*

The characteristics of contaminants affecting electrokinetic remediation are type, concentration, mobility, and chemical form. The physico-chemical interaction between soil media and contaminants depends on the types of contaminants such as inorganics (heavy metals, metalloid, and radionuclides) and organics (PAHs, pesticides, herbicides, etc.). Type of contaminant also determines whether the main mechanism for transport is electromigration, electroosmosis, or coupled effect of both. In the case of heavy metal contaminants, for example, electromigration and electroosmosis play a simultaneous role in transporting and removing the contaminants under normal conditions of soils. However, non-polar organic contaminants are best removed by electroosmotic flushing rather than electromigrative transport, without considering the oxidation effect induced by the electric field. The concentration of contaminants influences the efficacy of the process as well. Removal and current efficiencies can be decreased if the concentrations of non-target contaminants are higher than those of target contaminants. In order to explain this phenomenon, the transport (transference) number is introduced. The transport number (t_i) of the contaminant i is given as (Alshawabkeh, 1994; Kim, 2001):

$$t_i = (z_i u_i^* C_i) / \sum (z_i u_i^* C_i) \tag{5.13}$$

where z_i, u_i^*, and C_i are the charge, the effective ionic mobility, and the aqueous concentration of the contaminant i, respectively. The transport number gives the contribution of the i-th on to the total effective conductivity. The summation of transport numbers of all ions in the soil pore fluid should be equal to one. Equation (5.13) formalizes the dependence of the transport number of an individual ion on its effective ionic mobility, concentration, and total electrolyte concentration in the pore fluid. The transport number of a species will increase as the ionic concentration of that specific species increases. It indicates that as the concentration of a species decreases relative to the total electrolyte concentration in the pore fluid, its transport and removal under an electric field will be less efficient. Therefore, it is reasonable to assume that the efficiency of removal of a specific contaminant will decrease in time as its concentration in the pore fluid decreases (Alshawabkeh, 1994). In addition to the concentration of contaminants, their chemical form (or speciation) affects the efficiency of the process because it is directly related to the mobility and solubility of contaminants. Depending on the conditions of the surrounding environment, the contaminants are partitioned into various forms, such as (1) a dissolved fraction in pore fluid, (2) a water soluble or exchangeable fraction, (3) a specifically adsorbed fraction, (4) a precipitated fraction as insoluble carbonates, sulfides, phosphates, and oxides, (5) an organically complexed fraction, (6) a crystalline (hydro)oxides fraction, and (7) a residual fraction (Kim *et al.*, 2009a).

Even though the concentrations of contaminants are identical, the overall performance of the electrokinetic process appears to be totally different depending on their fractionation in soils. If contaminants exist as loosely bound fractions such as (1)–(4), for example, they tend to be relatively easily removed by the process. On the contrary, contaminants associated with organics or in crystal lattices such as (5)–(7) cannot be effectively removed or separated from soils (Kim et al., 2009a).

5.3.1.3   Voltage and current level
The electric field strength affects electromigration and electroosmosis, as shown in Equations (5.3) and (5.9). The common range of the electric field strength frequently used is 1–$100\,\text{V}\,\text{m}^{-1}$ (Alshawabkeh et al., 1999; Page and Page, 2002; Virkutyte et al., 2002). The level of electric current used in most implementation is in the order of a few tens of milliamperes per square centimeter. The electric current intensities control the rate of electrolysis of water through the reaction of Equations (5.14) and (5.15) (Hamed et al., 1991; Kim and Kim, 2002). The rate of H^+ generation (R_H) can be related to the rate of water electrolysis (R_w) in the anode based on Equation (5.11):

$$R_H\ [\text{mole s}^{-1}] = 2 \times R_w = 2 \times (I/2F) = I/F \qquad (5.14)$$

where I (A) is the current intensity and F is the Faraday constant [$96,485$ (A s) mole^{-1}]. Similarly, the rate of OH$^-$ generation (R_{OH}) can be calculated by the rate of water electrolysis (R_w) in the cathode based on Equation (5.15):

$$R_{OH}^-\ [\text{mole s}^{-1}] = R_w = I/F \qquad (5.15)$$

Although a high current intensity can generate more acid and increase the rate of transport to facilitate the contaminant removal process, it increases power consumption tremendously as power consumption is proportional to the square of electric current. An electric current density in the range of 1–$10\,\text{A}\,\text{m}^{-2}$ has been demonstrated to be the most efficient for the process (Alshawabkeh et al., 1999; Page and Page, 2002; Virkutyte et al., 2002). An optimum electric field strength and electric current density should be evaluated based on soil properties, electrode distance, and time requirements of the process. Details will be discussed in later sections.

5.3.2   Practical consideration for optimization of operation and design of electrokinetic remediation

5.3.2.1   Electrode
When one designs the electrodes for the electrokinetic process, the distance between anode and cathode is a crucial parameter because it controls the electric field strength applied and the duration of the process. The longer the distance considered, the fewer the number of electrodes required. Accordingly, the cost for production and installation of electrodes can be saved. On the other hand, the increase in time required by the process cannot be avoided, and the cost of operation must be increased. Therefore, duration of the process and cost should be considered when the distance between electrodes is determined. The time required by the process is affected by the velocity of contaminant transport as well as the distance between electrodes, and one must know the relation between time requirements and electrode distance when designing the appropriate distance of electrodes. The velocity, V [m s^{-1}], of contaminant transport under the applied electric field, E [V m^{-1}], can be calculated by electromigrative velocity, V_{em} [m s^{-1}], in Equation (5.3) and electroosmotic velocity, V_{eo} [m s^{-1}], in Equation (5.4):

$$V = V_{em} + V_{eo} \qquad (5.16)$$

Substituting Equations (5.2–5.4), and (5.10) into Equation (5.16):

$$V = (n\tau u_i + k_{eo})E \qquad (5.17)$$

If L is the distance of electrodes, then the time requirement (t) of the process is expressed as:

$$t = L/V = L/((n\tau u_i + k_{eo})E) \tag{5.18}$$

From a practical perspective, Equation (5.18) is not proper because the interaction between soil particles and contaminants is not considered. The transport of contaminants in soil media is delayed due to interactions such as sorption and precipitation. Therefore, the retardation (delaying) factor (R_d) should be taken into account:

$$t = (R_d L)/((n\tau u_i + k_{eo})E) \tag{5.19}$$

The retardation factor is influenced by characteristics of the soil and contaminant and should be evaluated prior to implementation of the process. As shown in Equation (5.19), there are three parameters (t, L, and E) to be determined when the electrokinetic process is designed, and one can be optimized by changes in the other two.

In addition to the distance between electrodes, electrode material is important. Chemically inert and electrically conducting materials such as graphite, coated titanium stainless steel, or platinum have been usually used as anodes to prevent dissolution of the electrode and generation of undesirable corrosion products in an acidic environment. Any conductive materials that do not corrode in a basic environment can be used as cathodes. When one selects the material for electrodes, the following aspects must be considered (Alshawabkeh *et al.*, 1999):

• Chemically inert and electrically conducting material.
• Availability of the material.
• Easy fabrication to the form required for the process.
• Easy installation in field.
• Costs for material, production or fabrication, and installation.

Another consideration for electrode design is the spacing between electrodes with the same polarity. The spacing of electrodes is subjected to the configuration of electrodes. Overall, there are two kinds of electrode arrangements applied: one-dimensional (1-D) and two-dimensional (2-D) arrays. In order to design an effective and efficient configuration of electrodes, several aspects must be considered:

• Electrically effective and ineffective spots or areas.
• Number and cost of electrodes required per unit area.
• Processing time required.

Figure 5.3 shows representative 1-D and 2-D electrode configurations. The 1-D electrode configuration is the easiest method to install, and the electric field is produced linearly. Depending on the spacing of electrodes, spots or areas of inactive (ineffective) electric field can be varied in this configuration. However, there are drawbacks to this configuration. Because the numbers of anode and cathode are identical, it is difficult to collect wastewater that contains contaminants from the effluent systems of electrode compartments. Additionally, more electrodes are required, compared with other configurations. To overcome the drawbacks of the 1-D configuration, the 2-D arrangement can be considered. Triangular, square, or hexagonal electrode configurations can be used for 2-D implementation. Depending on the type of contaminants targeted, the place of anode or cathode is determined. When cationic contaminants, such as heavy metals, are removed by the electrokinetic process, the cathode is placed at the center, and the anodes are placed on the perimeter to maximize the spread of the acidic environment generated by the anodes and to minimize the extent of the basic environment generated by the cathode. The 2-D configurations of electrodes generate nonlinear electric fields, that is, the electric potential gradient and electric current density gradually increases towards the center of the configuration, and the transport of contaminants is enhanced as the processing time increases. Compared with the 1-D array, the collection of wastewater is easy due to one effluent system being placed at the center. The numbers

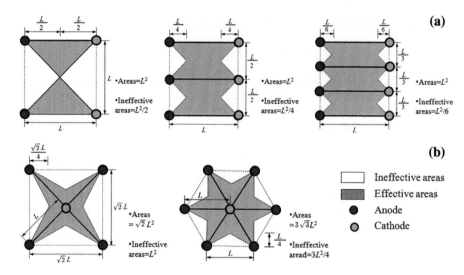

Figure 5.3. Typical configurations of electrodes. (a) one-dimensional configuration and (b) two-dimensional configuration (Alshawabkeh *et al.*, 1999b).

Table 5.3. Impact of electrode configuration on electrode requirements and size of ineffective areas[1].

Configuration		Electrode distance	spacing	No. of electrodes per cell (α)	Area of cell	No. of electrodes per unit area N	% increase
1-D		L	L	1	L^2	$1/L^2$	0
		L	$L/2$	2	L^2	$2/L^2$	100
		L	$L/3$	3	L^2	$3/L^2$	200
2-D	Square	R	$2^{1/2}R$	2	R^2	$1/R^2$	0
	Hexagonal	R	R	3	$(3^{3/2}R^2)/2$	$2/(3^{1/2}R^2)$	15.5

[1]From Alshawabkeh *et al.* (1999b).

of electrodes required can be estimated depending on the configuration, distance, and spacing of electrodes.

The number (N) of electrodes per unit surface area of the site to be treated (Alshawabkeh *et al.*, 1999b):

$$N = (\alpha/L^2)_{1\text{-}D} = (\alpha/\pi R^2)_{2\text{-}D} \tag{5.20}$$

where α is the number of electrodes per cell, and L and R are the distances of electrodes within 1-D and 2-D configurations, respectively. Representative values for α, L, and R are given in Table 5.3. Refer to Turer and Genc (2005) and Almeira *et al.* (2009) for more detailed information about the impact of different electrode configurations on the efficiency of electrokinetic remediation.

The shape of electrodes is important because it determines the surface area of the electrodes and affects the reactions occurring on the electrode surface. The common shapes used frequently are plate, rod, net, and cylinder.

5.3.2.2 *Electrolyte chemistry and enhancement scheme*

As discussed in Section 5.3.1.2, contaminants exist in different chemical forms in soil media depending on environmental conditions. Among those different forms, only dissolved fractions of contaminants are mobile and can be removed by the electrokinetic process. Therefore, other forms of contaminants should be transformed to their soluble and dissolved forms. However, the transformation processes in which the chemical forms of contaminants are changed are considered to be contaminant specific, reversible, and dependent on environmental conditions. The acidic condition is favorable in desorption and mobilization of cationic contaminants, such as heavy metals, while the basic environment aids in solubilization of anionic contaminants such as arsenite, arsenate, and fluoride. Therefore, the pH and composition of the electrolyte solution must be adjusted because the transformation processes of contaminants in the soil can be improved or hindered by electrolyte chemistry.

A variety of enhancement schemes have been proposed and tested to facilitate electrokinetic extraction of contaminants, as briefly addressed in Section 5.2.3.3. Normally, enhancement schemes can be embodied by injecting reagents, called as enhancing or enhancement agents, into the soil or electrolyte solutions. The enhancement agents are needed to control the soil chemistry and to promote solubilization (mobilization) of contaminants, and finally improve the overall efficiency of the process. Alshawabkeh *et al.* (1999) summarized the important characteristics of enhancement agents:

- They should not form insoluble salts with the contaminant within the range of pH values expected to develop during the process.
- They should form soluble complexes with the contaminant that can electromigrate efficiently under the electric field applied.
- They should be chemically stable over a wide range of pH values.
- They should have a higher affinity for the contaminant than the soil particle surface.
- They and the resulting complexes should not have a strong affinity for the soil particle surface.
- They should not generate toxic residue in the treated soil.
- They should not generate an excessive quantity of wastewater or the end products of the treatment process should be amenable to concentration and precipitation after use.
- They should be cost-effective including reagent cost and treatment costs for the waste collected and/or wastewater generated.
- They should not induce excessive solubilization of soil minerals or increase the concentrations of any regulated species in the soil pore fluid.
- If possible, they should complex with the target species selectively.

The enhancement schemes that have been most frequently used include enhancement agents for catholyte neutralization, membrane enhancement, and chelating or complexing agents. The common chelating or complexing agents tested are citric acid, EDTA, and surfactants. However, special caution should be exercised when enhancement schemes are considered because some agents significantly affect the zeta potential of the soil surface, and in turn, the direction of electroosmosis is changed (Eykholt and Daniel, 1994; Kim, 2001; Kim *et al.*, 2002a). The overall efficiency of electrokinetics can be influenced by unexpected variation of zeta potential and electroosmosis.

5.3.2.3 *Type of electricity*

The selection of a suitable type of electricity is important to improve the performance of the electrokinetic process. Figure 5.4 shows typical types of electricity that can be applied for electrokinetic remediation. The full-wave DC has usually been applied in electrokinetic removal of inorganic contaminants, such as heavy metals, arsenic species, and radionuclides. For the sake of extracting organic contaminants, however, AC can be used for the electrokinetic process because

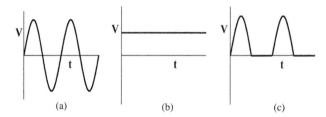

Figure 5.4. Types of electricity. (a) alternating current (AC), (b) full-wave direct current (DC), and (c) half-wave direct current (DC).

the contaminants are removed by the mechanism of soil heating rather than the transport mechanism. Compared with the full-wave DC, the half-wave DC can facilitate the transformation process of the contaminants, and promote solubilization or mobilization of contaminants due to its pulse effect. If contaminants initially exist as strongly bound fractions, the half-wave DC is favorable at the early stage of the process. After the application of the half-wave DC, the full-wave DC is supplied to transport the dissolved or soluble forms of contaminants. Except for three types of electricity, pulsed electricity has been applied to enhance the electrokinetic process (Ryu et al., 2009). Pulsed electricity can improve desorption and dissolution of contaminants and finally increase the overall efficacy of the process. Depending on the chemical forms of contaminants and the electrical features of the soil media, electricity can be optimized by selection of sole type or coupling of different types.

5.4   FIELD APPLICATIONS OF ELECTROKINETIC REMEDIATION

A number of studies on electrokinetic remediation have been undertaken by researchers using various contaminated sources, enhancement schemes, and monitoring and assessment techniques. The representative research is presented in this section, which focuses on electrokinetic remediation of soils contaminated with inorganic contaminants, such as metals, heavy metals, arsenic, and fluorine. Prior to introducing field applications, the laboratory- or pilot-scale studies performed over the past decade are summarized (Table 5.4). The main topics of research are electrokinetic removal of contaminants from various types of soils, sediments, tailings, and sludges. In addition, several enhancement schemes have been evaluated to improve the performance of electrokinetic remediation because the most important aspect in the removal of heavy metals from a solid matrix is their mobilization. Various enhancing reagents have been tested on electrolyte or pretreatment solutions, and some elucidated different effects and trends depending on the different metal species, even under equivalent conditions. Reddy and his colleagues conducted a number of investigations to evaluate many chemical reagents, such as acetic acid, citric acid, EDTA (ethylenediaminetetraacetic acid), DTPA (diethylene triamine penta acetic acid), KI (potassium iodide), HPCD (hydroxypropyl-β-cyclodextrin), H_2SO_4, NaOH and NaCl as electrolyte solutions, and humic acid, ferrous iron and sulfide as reducing agents, to remove mixed heavy metals (Cr, Ni and Cd) from kaolin and glacial till soils (Al-Hamdan and Reddy, 2006; Reddy and Ala, 2005; Reddy and Chinthamreddy, 2003; Reddy et al., 2001a; 2001b; 2004). The technique using ion exchange membranes for electrokinetics is referred to as electrodialytic remediation. When an electrokinetic treatment is applied without any conditioning, metals may precipitate as hydroxides near the cathode region where the pH is increased, resulting in a decreased removal efficiency. A popular enhancement scheme to prevent the hydroxide precipitates is to apply an ion exchange membrane to the electrode compartments (Gardner et al., 2007; Hansen et al., 1997; Kim et al., 2005a; Li et al., 1998; Nystroem et al., 2005; Ottosen et al., 2003; Pedersen et al., 2005; Ribeiro et al., 2000). Many researchers have suggested numerous techniques to predict and evaluate process efficiency as well. Because the initial chemical forms of

Table 5.4. Representative studies on electrokinetic remediation of soils contaminated with inorganics over the last decade.

Media	Contaminant	Pollution	Note	Reference
Shooting range soil (sandy loam)	Cu, Pb, Zn	Polluted	Evaluation of removal efficiencies for 5 different enhancement schemes: acidic electrokinetics, acidic bioelectrokinetics, EDTA electrokinetics, EDTA bioelectrokinetics, and acidic + neutral electrokinetics	Lee and Kim (2010)
Factory site (sandy)	Cr(V)	Polluted	Comparison of 3 different electrode arrangements: the highest performance in 2D crossed electric field	Zhang et al. (2010)
Agricultural area (sandy loam)	Cd, Ni, Zn	Spiked	Evaluation of the efficiency of three kinds of chelate agents: nitrilotriacetic acid (NTA), diethylenetriaminepentaacetic acid (DTPA), and diaminocycloexsanetetraacetic acid (DCyTA)	Giannis et al. (2010)
Kaolinite, and clay liner (silty clay)	Zn	Spiked	Investigation of the effect of calcite or carbonate ($CaCO_3$) on electrokinetic removal of Zn: a decrease in the rate of heavy metal removal with an increase in the carbonate quantity	Ouhadi et al. (2010)
Dredged sediment	As, Cd, Cr, Cu, Ni, Pb, Zn, PAH	Polluted	Evaluation of electrokinetic removal of heavy metals and PAH under different operational conditions using different chelating agents such as EDTA and surfactant	Colacico et al. (2010)
Tailing (silt)	As, Ca, Cd, Co, Cu, Mo, Ni, Pb, Sr	Polluted	Study on enhanced electrokinetics using alkaline agents such as ammonium oxalate and sodium hydroxides; evaluation of chemical forms of contaminant using sequential extraction	Isosaari and Sillanpää (2010)
Tailing (sandy loam)	As	Polluted	Investigation of a hybrid method integrating anaerobic bioleaching and electrokinetics for removal of As	Lee et al. (2009)
Field soil (silty loam)	F	Polluted	Evaluation of the feasibility of anolyte conditioning on electrokinetic remediation of fluorine-contaminated soil	Kim et al. (2009b)
Power plant soil (silty loam)	Ni, Zn	Polluted	Investigation of the feasibility of catholyte conditioning with the acidic solution and pre-treatment of soil with acidic solution for the electrokinetic remediation of Zn and Ni contaminated field soil	Kim et al. (2009c)
Kaolinite glacial till (clay)	Cr, Ni, Cd	Spiked	Different removal efficiency; kaolinite > glacial till; geochemical assessment using MINEQL+	Al-Hamdan and Reddy (2008)
Incinerator bottom ash	Pb, Cu, Zn, Cd	Polluted	Comparative experiments by different current densities and process durations	Traina et al. (2007)
Estuarine harbor sediment	As, Cd, Cr, Cu, Ni, Pb, Zn	Polluted	Electrodialytic remediation (EDR); average removal efficiency: 75%; dissolved organic carbon (DOC) from sediment: before EDR < after EDR	Gardner et al. (2007)

(Continued)

Table 5.4. Continued.

Media	Contaminant	Pollution	Note	Reference
Black cotton soil	Cr, Fe	Spiked	Electroosmotic flow monitoring; different removal trends by heavy metals	Sivapullaiah et al. (2007)
Waste water sludge	Cr, Cu, Fe, Ni, Pb, Zn	Polluted	Comparison by different electrolyte solutions; highest efficiency: citric acid; Cu > Pb > Ni > Fe > Zn > Cr	Yuan and Weng (2006)
Kaolinite	Cr(VI), Cr(III), Ni, Cd	Spiked	Adsorption and precipitation properties of single and multiple heavy metals	Al-Hamdan and Reddy (2006)
Kaolinite, tailing	As	Spiked, polluted	Comparison of the effectiveness of two kinds of enhancing agents (KH_2PO_4 and NaOH); evaluation on factors affecting the efficiency of the electrokinetic process in removing As contaminants from soils	Kim et al. (2005b)
Kaolinite	Pb, Cd	Spiked	Enhanced by ion exchange membrane; efficiency evaluation depending on electrode configuration	Kim et al. (2005a)
Red soil	Cu, Zn	Polluted	Catholyte pH controlled by lactic acid and $CaCl_2$; without $CaCl_2$: Zn > Cu; with $CaCl_2$: only enhanced Cu	Zhou et al. (2005)
Silt clay	Pb, Zn, Cu	Spiked	Comparison of single and multiple heavy metals; two types of electrode geometry	Turer and Genc (2005)
Field soil (clay)	19 metals (spiked: Pb, Hg)	Polluted, spiked	Enhanced by different extracting solutions; efficient solutions: EDTA, KI	Reddy and Ala (2005)
Kaolinite	Cr(VI), Cr(III), Ni, Cd	Spiked	Evaluation of pH dependent adsorption of metals; zeta potential change by heavy metals	Al-Hamdan and Reddy (2005)
Waste water sludge	As, Cr, Ni, Pb, Cu, Zn	Polluted	Sludge pre-acidification & catholyte pH control; sequential extraction after treatment; Cu > Zn > Ni > Cr > As > Pb	Wang et al. (2005)
Harbor sediment	Cu, Zn, Pb Cd	Polluted	Electrodialytic remediation; control pH & liquid-solid ratio; maximum removal efficiency: 87% Cu, 98% Cd, 97% Zn, 96% Pb	Nystroem et al. (2005)
Incineration fly ash	Cd, Pb, Zn, Cu, Cr	Polluted	Electrodialytic remediation; enhanced by ammonium citrate filling solution; 86% Cd, 20% Pb, 62% Zn, 81% Cu, 44% Cr	Pedersen et al. (2005)
Kaolinite	Cr, Ni, Cd	Spiked	EDTA-enhanced electrokinetics; high removal efficiency of only Cr in high pH	Reddy et al. (2004)

Soil/Site	Metals	Spiked/Polluted	Description	Reference
Glacial till (clay)	Cr, Ni, Cd	Spiked	Enhanced by various electrolyte solutions; simultaneous removal of multiple heavy metals: NaCl/EDTA or acetic acid	Reddy and Chinthamreddy (2004)
Waste disposal site	Cr, Cd	Polluted	Different removal efficiency depending on scale; bench scale < field scale; field scale: 78% Cr, 70% Cd after 6 months	Gent et al. (2004)
Kaolinite	Cr, Ni, Cd	Spiked	Sequentially enhanced electrokinetics by electrolyte conditioning	Reddy and Chinthamreddy (2003)
Factory site (clay)	Ca, Mg, Mn, Fe, Ni, Cu, Zn, Pb, Cd	Polluted	Selective leaching test after electrokinetics; low potential for remobilization of metals after electrokinetics	Suèr et al. (2003)
Factory site	Cu, Cd	Polluted	Combination of electrokinetics and phytoremediation	O'Connor et al. (2003)
Tailing (mainly silt)	Cd, Cu, Pb, Zn	Polluted	Efficiency evaluation depending on heavy metal speciation in soil; main removal mechanism: electromigration > electroosmosis	Kim et al. (2002a)
Various natural clay	Cu, Pb, Zn	Spiked	Heavy metal migration dominated by crystalline clay minerals; lowest efficiency: humic-allophanic & allophonic soils	Darmawan and Wada (2002)
Waste water sludge	Cd, Cr, Cu, Pb	Polluted	Pilot scale study; abiotic and biotic speciation of heavy metals	Kim et al. (2002b)
Sand, clay	Pb, Zn, Cd	Spiked	pH-dependent removal efficiency: Cd and Zn > Pb, sand > clay	Vengris et al. (2001)
Kaolinite, glacial till (clay)	Cr, Ni, Cd	Spiked	Evaluation of effects of multiple heavy metals; removal efficiency depending on soil pH, polarity of contaminants, type of soil	Reddy et al. (2001a)
Kaolinite, glacial till (clay)	Cr(VI), Cr(III), Ni, Cd	Spiked	Different migration of initial metal species; efficiency evaluation by sequential extraction	Reddy et al. (2001b)
Kaolinite, tailing (clay)	Pb, Cd	Spiked, polluted	Different removal efficiency: Pb < Cd, kaolinite > tailing	Kim et al. (2001)

contaminants in soils are crucial in electrokinetic remediation, as discussed in Section 5.3.1.2, researchers have investigated their speciations using sequential extraction or selective leaching techniques prior to applying the electrokinetic process. In addition, the stability of residual contaminants in soil has been assessed after electrokinetic remediation. Based on the results of the lab- or bench-scale feasibility studies presented so far, the electrokinetics has been established as a promising technology for remediation of soils, especially fine-grained ones in which other remediating technologies, such as soil washing, have failed.

The first field-scale application of electrokinetic soil remediation was carried out by Geokinetics in 1987 (Lageman, 1993). Subsequently, a number of field works have been performed by many researchers and companies, focusing on the practical field-scale implementation of electrokinetic remediation of soils contaminated with inorganics such as heavy metals. The field-scale feasibility study of electrokinetic remediation by Banerjee et al. (1989), undertaken at a Superfund site at Corvallis, Oregon, was one among the other early field-scale studies (Oonnittan et al., 2009). Table 5.5 summarizes the field-scale projects of electrokinetic remediation undertaken to date. Field-scale electrokinetic remediation can be usually implemented by the application of low-level direct current (DC) between electrodes placed in a contaminated area. A variety of conditions and parameters of the process should be optimized prior to field works through laboratory tests, numerical methods, and/or design procedures. The processes adopted at each site differ in one or more aspects. Basically, two approaches are defined depending on the type of contaminant (Oonnittan et al., 2009). The first approach is enhanced removal in which the contaminants are transported by electromigration and/or by electroosmosis toward the electrodes for subsequent removal. This approach is applicable for the removal of inorganic contaminants such as heavy metals. The second approach is treatment without removal, which involves the electroosmotic transport of contaminants through the treatment zones and may also include the frequent reversal of polarity of electrodes to control the direction of contaminant movement. This approach was developed for the removal of organic contaminants from soils (USEPA, 1997). Although there have been some major hindrances during field operations undertaken so far, some of the projects shown in Table 5.5 were able to meet their remediation goals. Results of the field implementations demonstrate that electrokinetics is a promising technology for the effective remediation of soils contaminated by various contaminants. Nevertheless, the field applications that have been undertaken reveal there are several factors, such as soil heterogeneities, remediation time, formation of by-products, and soil saturation that limit or affect the overall performance of the technology in the field. In order to improve the effectiveness of the process, a detailed study of each case is needed to establish the most adequate operating conditions, including current intensity or electric field strength, electrode deposition, and chemical conditioning of electrolyte solutions. Recently, several demonstration projects have been undertaken to remediate various types of sediments. For example, a field study was carried out for an alternative sediment remediation technology known as electrochemical geo-oxidation (ECGO) in Duluth, Minnesota, USA, from 2002 to 2007 and in Copenhagen, Denmark in 2006, for removal of PAHs, PCBs, Hg, and other miscellaneous contaminants from dredged harbor sediments (Wittle et al., 2009). In addition, the demonstration of field-scale electrokinetics coupled with permeable reactive barriers (PRBs) was undertaken at a landfill site to clean up the groundwater as well as soil contaminated by the uncontrolled release of landfill leachate (Chung and Lee, 2007). In summary, a number of field-scale projects have demonstrated the effectiveness of electrokinetic remediation for removal of various types of contaminants from different media such as soil, sediment, and groundwater. However, the field demonstrations indicate that detailed study of contaminated sites is prerequisite for the successful application of the technology. One must consider several basic questions (Lageman and Pool, 2009): What parameters must be known to calculate the necessary energy and time to reach a certain remediation goal? How are these parameters obtained and what other field information is needed? What field equipment should be used? Finally, how is an electrokinetic project designed and operated?

Table 5.5. Major field-scale applications of electrokinetic remediation of soils contaminated with inorganics.

Site description	Contaminant and concentration	Scheme	Note	Reference
Former iron refinery plant site (Janghang, Korea)	As: 52–79 mg kg^{-1} Cu: 62–124 mg kg^{-1} Pb: 111–204 mg kg^{-1}	Electrokinetics and Bio-EK (electrokinetics coupled with bioleaching)	Treated zone: silty clay loam with a volume of 150 m^3 (0.8 m of depth); demonstrating the improvement in removal efficiency using bioleaching prior to EK: As (32.30%), Cu (49.77%), and Pb (52.88%) by EK; As (64.04%), Cu (64.88%), and Pb (56.95%) by EK	Halla Engrg. & Construction, Corp. (2011)
Former galvanizing plant site (The Hague, The Netherlands)	Zn: 2000 μg L^{-1}, in groundwater	In-situ remediation using electrostimulated groundwater extraction (ESGE)	Treated zone: medium-fine sand and silty clay layer with a volume of 5800 m^3; average Zn concentration of <600 μg L^{-1} after treatment of 2 years	Lageman and Pool (2009)
Operational galvanizing plant site (Heerenberg, The Netherlands)	Ni: 1350 mg kg^{-1} Zn: 1300 mg kg^{-1} Ni: 3500 μg L^{-1} in groundwater	In-situ electrokinetics	Treated zone: sandy loam and silty sand with a volume of 4300 m^3; applied to the depth 6 m below ground surface; Ni (15 mg/kg), Zn (75 mg/kg in soil, 15 μg L^{-1} in groundwater) after treatment	Lageman and Pool (2009)
Former gasworks site (Oostburg, The Netherlands)	CN: 930 mg kg^{-1}	In-situ electrokinetics	Treated zone: clay and sandy clay soil with a volume of 120 m^3; applied to the depth 4 m below ground surface; Ni (15 mg kg^{-1}), Zn (75 mg kg^{-1} in soil, 15 μg L^{-1} in groundwater) after treatment of 3 months	Lageman and Pool (2009)
Naval Air Weapons Station (NAWS) (Point Mugu, CA)	Cd: 5–20 mg kg^{-1} Cr: 180–1100 mg kg^{-1}	In-situ electrokinetics	Treated zone: a volume of 120 m^3 with the depth of 3 m; citric acid used to maintain the cathode pH at 4; after 6 months of treatment, 78% of the soil volume had been cleared of chromium or treated to below natural background levels; 70% of the soil between the electrodes had been cleared of cadmium contamination	Gent et al. (2004)
Battery shop site and intermediate maintenance facility site (Honolulu, HA)	Pb: 82,230–8270 mg kg^{-1}	Ex-situ electrokinetics	The innovative feature by using EDTA as a chelating agent in the electrolyte solution; the effective removal of Pb from polluted soils with no generation of harmful wastes	USEPA (2003)
Naval Air Weapons Station (NAWS) (Point Mugu, CA)	Cd: 1810 mg kg^{-1} Cr: 25,100 mg kg^{-1}	In-situ electrokinetics	The aim to reduce the contaminant concentrations below the regulatory action levels for metal concentration and toxicity criteria; usage of citric acid as an amendment to control the formation of the pH front in the treatment area	USAEC (2000)

(Continued)

Table 5.5. Continued.

Site description	Contaminant and concentration	Scheme	Note	Reference
Chemical waste landfill site (Albuquerque, NM)	Cr	In-situ electrokinetic extraction (ISEE)	The ISEE developed by Sandia National Laboratories (SNL) to remove hexavalent Cr from unsaturated soil; applied to the depth of 4.3 m; circulation of anode fluid for chemical conditioning and pH control	USEPA (1998)
Military airbase (Woensdrecht, The Netherlands)	Cd: 660 mg kg^{-1}; Cu: 770 mg kg^{-1}; Cr: 7300 mg kg^{-1}; Ni: 860 mg kg^{-1}; Pb: 730 mg kg^{-1}; Zn: 2600 mg kg^{-1}	Ex-situ electrokinetics	Treated zone: clay soil with a volume of 2500 m^3; undertaken to reduce Cd to <50 mg kg^{-1} after treatment of 2 years (1992–1994)	Lageman (1993)
Temporary landfill site (Stadskanaal, The Netherlands)	Cd: 20–3400 (>180) mg kg^{-1} (also CN, Pb, Zn)	Ex-situ electrokinetics	Treated zone: argillaceous soil with a volume of 2500 m^3; undertaken to reduce Cd to <10 mg kg^{-1} after treatment of 2 years (1990–1992)	Lageman (1993)
Former timber impregnation plant site (Loppersum, The Netherlands)	As: 400–500 (>250) mg kg^{-1}	In-situ electrokinetics	Treated zone: heavy clay soil with a volume of 300 m^3; reduction of As concentration to <30 mg kg^{-1} after treatment of 65 days (1989)	Lageman (1993)
Galvanizing plant site (Delft, The Netherlands)	Zn: 2410 (>1400) mg kg^{-1}	In-situ electrokinetics	Treated zone: clay soil with a volume of 250 m^3; reduction of Zn concentration to 600 mg kg^{-1} after treatment of 8 weeks (1987)	Lageman (1993)
Former paint factory site (Groningen, The Netherlands)	Cu: >5000 mg kg^{-1}; Pb: >500–1000 mg kg^{-1}	In-situ electrokinetics	First pilot project on electrokinetic remediation; treated zone: peat and clay soil with a volume of 300 m^3; reduction of Cu and Pb concentrations to <200 and <280 mg kg^{-1}, respectively, after treatment of 8 weeks (1987)	Lageman (1993)
Superfund site (Corvallis, Oregon)	Cr	In-situ electrokinetics	Field study to evaluate the technical feasibility of electrokinetic remediation technique to treat a site contaminated with Cr plating wastes	Banerjee et al. (1989)

5.5   PROSPECTS FOR ELECTROKINETIC REMEDIATION

The soil remediation technologies developed so far can be classified into two groups depending on the stage of development. The first generation (1G) technologies include *ex-situ* ones, such as soil washing, land farming, incineration, and compositing. These technologies are considered to be economically ineffective, even though they have shown a relatively higher performance. The next generation (2G) technologies comprise *in-situ* ones, such as soil flushing, soil vapor extraction, and electrokinetics. The 2G technologies are known to be effective for the removal of contaminants and are economically viable. However, their major drawback is that their overall performance is totally dependent on the type and concentration of contaminant. Especially, the sole (individual) *in-situ* technology cannot be effectively operated in removing multiple (mixed) contaminants, such as multiple heavy metals, multiple organics, or multiple metals and organics. To overcome the drawbacks of 2G technologies and to create new synergetic effects, third generation (3G) technologies have recently been developed and investigated through integration and/or coupling of different individual technologies, for example, bioremediation/SS, soil flushing/phytoremediation, and so on. The major advantage of 3G technologies is that they are independent of the type and concentration of contaminant, and therefore, can be used effectively to remove mixed contaminants in various levels of concentration in soil. Integrated 3G technologies are expected to be more developed in the future, and they may be the key solution for remediation of contaminated soils. From this point of view, the electrokientic remediation should be coupled with other technologies to improve its performance and overcome its main defects. For example, electrokinetics incorporated with bioremediation can effectively remediate soils contaminated with multiple contaminants such as inorganics and organics. When determining the sequence of integration of different technologies, one must consider the properties of soils and contaminants. The 3G technologies related to electrokinetics, which are currently ongoing and also expected to come into spotlight in the future, are exemplified as follows:

- Electroheated extraction.
- Electrokinetic biobarrier.
- Electrokinetic chemical oxidation/reduction.
- Electrokinetic permeable reactive barrier (PRB).
- Electrokinetic fence.
- Electrokinetic stabilization.
- Electrokinetic chemical oxidation/reduction.
- Electrokinetic phytoremediation.
- Electrokinetic bioremediation (or bioelectric remediation).

Through integration and/or coupling, the efficacy of electrokinetics could be extended to remediate various contaminated media including sediment and groundwater as well.

ACKNOWLEDGEMENTS

The authors thank Professor Ouhadi and two anonymous reviewers for their valuable and constructive comments. This work was supported by Dr. S.-O. Kim's 2012 sabbatical program funded by Gyeongsang National University.

REFERENCES

Acar, Y.B. & Alshawabkeh, A.N.: Principles of electrokinetic remediation. *Environ. Sci. Technol.* 27 (1993), pp. 2638–2647.
Al-Hamdan, A.Z. & Reddy, K.R.: Geochemical reconnaissance of heavy metals in kaolin after electrokinetic remediation. *J. Environ. Sci. Health* A41 (2006), pp. 17–33.

Al-Hamdan, A.Z. & Reddy, K.R.: Surface speciation modeling of heavy metals in kaolin: Implications for electrokinetic soil remediation processes. *Adsorption* 11 (2005), pp. 529–546.

Al-Hamdan, A.Z. & Reddy, K.R.: Transient behavior of heavy metals in soils during electrokinetic remediation. *Chemosphere* 71 (2008), pp. 860–871.

Almeira, J., Peng, C. & Wang, Z: Effect of different electrode configurations on the migration of copper ions during the electrokinetic remediation process. *Asia-Pacific J. Chem. Eng.* 4 (2009), pp. 581–585.

Alshawabkeh, A.N.: *Theoretical and experimental modeling of removing contaminants from soils by an electric field.* PhD Thesis, Department of Civil and Environmental Engineering, Louisiana State University and Agricultural and Mechanical College, Ann Arbor, MI, 1994.

Alshawabkeh, A.N. & Acar, Y.B.: Electrokinetic remediation: II. Theoretical model. *J. Geotech. Eng. ASCE* 122 (1996), pp. 186–196.

Alshawabkeh, A.N., Gale, R.J., Ozsu-Acar, E. & Bricka, R.M.: Optimization of 2-D electrode configuration for electrokinetic remediation. *J. Soil Contam.* 8 (1996), pp. 617–635.

Alshawabkeh, A.N., Yeung, A.T. & Bricka, M.R.: Practical aspects of *in-situ* electrokinetic extraction. *J. Environ. Eng.* 125 (1999), pp. 27–35.

Banerjee, S., Horng, J.J., Ferguson, J.F. & Nelson, P.O.: Field scale feasibility study of electrokinetic remediation. USEPA, Cincinnati, OH, 1989.

Chung, H.I. & Lee, M.H.: A new method for remedial treatment of contaminated clayey soils by electrokinetics coupled with permeable reactive barriers. *Electrochim. Acta* 52 (2007), pp. 3427–3431.

Colacicco, A., Gioannis, G.D., Muntoni, A., Pettinao, E. & Polettini, A.: Enhanced electrokinetic treatment of marine sediments contaminated by heavy metals and PAHs. *Chemosphere* 81 (2010), pp. 46–56.

Darmawan, R. & Wada, S.I.: Effect of clay mineralogy on the feasibility of electrokinetic soil decontamination technology. *Appl. Clay Sci.* 20 (2002), pp. 283–293.

DOE: Rapid Commercialization Initiative (RCI) – Final Report for an Integrated *In-situ* Remediation Technology (Lasanga™). United States Department of Energy Office of Environmental Management, Report Number DOE/OR/22459-1, Pducah, KY, 1998a.

DOE: Record of Decision for Remedial Action at Solid Waste Management Unit 91 of Waste Area Group 27 at the Paducah Gaseous Diffusion Plant. United States Department of Energy Office of Environmental Management, Report Number DOE/OR/06-157&D2, Pducah, KY, 1998b.

Dzombak, D.A. & Morel, F.M.M.: *Surface complexation modelling: Hydrous ferric oxide.* John Wiley & Sons, Inc., New Jersey, 1990.

Eykholt, G.R. & Daniel, D.E.: Impact of system chemistry on electroosmosis in contaminated soil. *J. Geotechnical Eng.* 129 (1994), pp. 797–815.

Gardner, K.H., Nystroem, G.M. & Aulisio, D.A.: Leaching properties of estuarine harbor sediment before and after electrodialytic remediation. *Environ. Eng. Sci.* 24 (2007), pp. 424–433.

Gent, D.B., Bricka, R.M., Alshawabkeh, A.N., Larson, S.L., Fabian, G. & Granade, S.: Bench and field-scale evaluation of chromium and cadmium extraction by electrokinetics. *J. Haz. Mater*. 110 (2004), pp. 53–62.

Giannis, A., Pentari, D., Wang, J.Y. & Gidarakos, E.: Application of sequential extraction analysis to electrokinetic remediation of cadmium, nickel and zinc from contaminated soils. *J. Hazard. Mater*. 184 (2010), pp. 547–554.

Halla Engineering and Construction, Corp.: A field demonstration for remediation of heavy metal contaminated-soil using integrated process of bioleaching and elctrokinetics. Final report on development of contamination remediation technology (in Korean). Project No. 173-081-027, Ministry of Environment, Korea, 2011.

Hamed, J., Acar, Y.B. & Gale, R: Pb(II) removal from kaolinite by electrokinetics. *J. Geotech. Eng.* 117 (1991), pp. 241–269.

Hansen, H.K., Ottosen, L.M., Kliem, B.K. & Villumsen, A.: Electrodialytic remediation of soils polluted with Cu, Cr, Hg, Pb and Zn. *J. Chem. Technol. Biotechnol.* 70 (1997), pp. 67–73.

Holmes, P.J.: *The electrochemistry of semiconductors.* Academic Press, London, UK, 1962.

Ho, S.V., Athmer, C.J., Sheridan, P.W., Hughes, B.M., Orth, R., McKenzie, D., Brodsky, P.H., Shapiro, A., Thornton, R., Salvo, J., Schultz, D., Landis, R., Griffith, R. & Shoemaker, S.: The Lasagna technology for in situ soil remediation. 1. Small field test. *Environ. Sci. Technol.* 33 (1999a), pp. 1086–1091.

Ho, S.V., Athmer, C., Sheridan, P.W., Hughes, B.M., Orth, R., McKenzie, D., Brodsky, P.H., Shapiro, A., Thornton, R., Salvo, J., Schultz, D., Landis, R., Griffith, R. & Shoemaker, S.: The Lasagna technology for in situ soil remediation. 2. Large field test. *Environ. Sci. Technol.* 33 (1999b), pp. 1092–1099.

Isosaari, P. & Sillanpää, M.: Electromigration of arsenic and co-existing metals in mine tailings. *Chemosphere* 81 (2010), pp. 1155–1158.

Kim, D.H., Jeon, C.S., Baek, K., Ko, S.H. & Yang, J.S.: Electrokinetic remediation of fluorine-contaminated soil: Conditioning of anolyte. *J. Hazard. Mater.* 161 (2009b), pp. 565–569.

Kim, D.H., Ryu, B.G., Park, S.W., Seo, C.I. & Baek, K.: Electrokinetic remediation of Zn and Ni-contaminated soil. *J. Hazard. Mater.* 165 (2009c), pp. 501–505.

Kim, K.W. & Kim, S.O.: Electrokinetic soil processing for energy from waste. In: V.I. Grover, V.K. Grove & W. Hogland (eds): *Recovering energy from waste: Various Aspects.* Science Publishers, Inc., New Hampshire, USA, 2002, pp. 107–140.

Kim, K.W., Lee, K.Y. & Kim, S.O.: Electrokinetic remediation of mixed metal contaminants. In: K.R. Reddy & C. Camesselle (eds): *Electrochemical remediation technologies for polluted soils, sediments and groundwater.* Science Publishers, Inc., New Hampshire, USA, 2009a, pp. 287–314.

Kim, S.O.: *Electrokinetic remediation of heavy metal contaminated soils and sludges.* PhD Thesis, Department of Environmental Sciences and Engineering, Gwangju Institute of Science and Technology (GIST), Gwangju, Republic of Korea, 2001.

Kim, S.O., Kim, J.J., Yun, S.T. & Kim, K.K.: Numerical and experimental studies on cadmium (II) transport in kaolinite clay under electrical fields. *Water Air Soil Poll.* 150 (2003), pp. 135–162.

Kim, S.O., Kim. K.W. & Stüben, D.: Evaluation of electrokinetic removal of heavy metals from tailing soils. *J. Environ. Eng., ASCE,* 128 (2002a), pp. 705–715.

Kim, S.O., Kim, W.S. & Kim, K.W.: Evaluation of electrokinetic remediation of arsenic-contaminated soils. *Environ. Geochem. Health* 27 (2005b), pp. 443–453.

Kim, S.O, Moon, S.H. & Kim, K.W.: Removal of heavy metals from soils using enhanced electrokinetic soil processing. *Water Air Soil Pollut.* 125 (2001), pp. 259–272.

Kim, S.O., Moon, S.H., Kim, K.W. & Yun, S.T.: Pilot scale study on the ex situ electrokinetic removal of heavy metals from municipal wastewater sludges. *Water Res.* 36 (2002b), pp. 4765–4774.

Kim, W.S., Kim, S.O. & Kim, K.W.: Enhanced electrokinetic extraction of heavy metals from soils assisted by ion exchange membranes. *J. Hazard. Mater.* B118 (2005a), pp. 93–102.

Kruyt, H.R.: *Colloid science (I): Irreversible systems.* Elsevier Publishing Co., Amsterdam, The Netherlands, 1952.

Lageman, R.: Electro-reclamation: Applications in The Netherlands. *Environ. Sci. Technol.* 27 (1993), pp. 2648–2650.

Lageman, R. & Pool, W.: Experience with field applications of electrokinetic remediation. In: K.R. Reddy & C. Camesselle (eds): *Electrochemical remediation technologies for polluted soils, sediments and groundwater.* Science Publishers, Inc., NH, 2009, pp. 697–717.

Lee, K.Y. & Kim, K.W: Heavy metal removal from shooting range soil by hybrid electrokinetics with bacteria and enhancing agents. *Environ. Sci. Technol.* 44 (2010), pp. 9482–9487.

Lee, K.Y., Yoon, I.H., Lee, B.T., Kim, S.O. & Kim, K.W: A novel combination of anaerobic bioleaching and electrokinetics for arsenic removal from mine tailing soil. *Environ. Sci. Technol.* 43 (2009), pp. 9354–9360.

Li, Z., Yu, J.W. & Neretnieks, I.: Electroremediation: Removal of heavy metals from soils by using cation selective membrane. *Environ. Sci. Technol.* 32 (1998), pp. 394–397.

Lorenz, P.B.: Surface conductance and electrokinetic properties of kaolinite beds. *Clays Clay Mineral* 17 (1969), pp. 223–231.

Mithcell, J.K.: Conduction phenomena: from theory to geotechnical practice. *Géotechnique* 41 (1991), pp. 299–340.

Mitchell, J.K.: *Fundamentals of soil behavior.* 2nd ed., John Wiley & Sons, Inc., New Jersey, 1993.

Mitchell, J.K. & Soga, K.: *Fundamentals of soil behavior.* 3rd ed., John Wiley & Sons, Inc., New Jersey, 2005.

Mitchell, J.K. & Yeung, T.C.: Electrokinetic flow barriers in compacted clay. *Transportation Research Records, No. 1289,* National Research Council, Washington DC, 1991.

Nystroem, G.M., Ottosen, L.M. & Villumsen, A.: Electrodialytic removal of Cu, Zn, Pb, and Cd from harbor sediment: Influence of changing experimental conditions. *Environ. Sci. Technol.* 39 (2005), pp. 2906–2911.

O'Connor, C.S., Lepp, N.W., Edwards, R. & Sunderland, G.: The combined use of electrokinetic remediation and phytoremediation to decontaminate metal-polluted soils: A laboratory-scale feasibility study. *Environ. Monit. Assess.* 84 (2003), pp. 141–158.

Oonnittan, A., Sillanpaa, M., Cameselle, C. & Reddy, K.R.: Field applications of electrokinetic remediation of soils contaminated with heavy metals. In: K.R. Reddy & C. Camesselle (eds): *Electrochemical remediation technologies for polluted soils, sediments and groundwater.* Science Publishers, Inc., NH, 2009, pp. 609–624.

Ottosen, L.M., Hansen, H.K. & Hansen, C.B.: Water splitting at ion-exchange membranes and potential differences in soil during electrodialytic soil remediation. *J. Appl. Electrochem.* 30 (2003), pp. 1199–1207.

Ouhadi, V.R., Yong, R.N., Shariatmadari, N., Saeidijam, S., Goodarzi, A.R. & Safari-Zanjani, M.: Impact of carbonate on the efficiency of heavy metal removal from kaolinite soil by the electrokinetic soil remediation method. *J. Hazard. Mater.* 173 (2010), pp. 87–94.

Page, M.M. & Page, C.L.: Electroremediation of contaminated soil. *J. Environ. Engrg.* 128 (2002), pp. 208–219.

Pamukcu, S. & Wittle, J.K.: Electrokinetically enhanced in situ soil decontamination. In: D.L. Wise & D.J. Trantolo (eds): *Remediation of hazardous waste contaminated soils*. Marcel Dekker, Inc., New York, 1994, pp. 245–298.

Pedersen, A.J., Ottosen, L.M. & Villumsen, A.: Electrodialytic removal of heavy metals from municipal solid waste incineration fly ash using ammonium citrate as assisting agent. *J. Hazard. Mater.* B122 (2005), pp. 103–109.

Probstein, R.F.: *Physicochemical hydrodynamics-An introduction*. 2nd ed., John Wiley & Sons, Inc., New Jersey, USA, 2003.

Reddy, K.R. & Ala, P.R.: Electrokinetic remediation of metal-contaminated field soil. *Separation Sci Technol* 40 (2005), pp. 1701–1720.

Reddy, K.R. & Cameselle, C.: Overview of electrochemical remediation technologies. In: K.R. Reddy & C. Cameselle (eds): *Electrochemical remediation technologies for polluted soils, sediments and groundwater*. Science Publishers, Inc., NH, 2009a, pp. 3–28.

Reddy, K.R. & Cameselle, C.: *Electrochemical remediation technologies for polluted soils, sediments and groundwater*. John Wiley & Sons, Inc., New Jersey, 2009b, pp. 195–284.

Reddy, K.R. & Chinthamreddy, S.: Sequentially enhanced electrokinetic remediation of heavy metals in low buffering clayey soils. *J. Geotech. Geoenviron. Eng.* 129 (2003), pp. 263–277.

Reddy, K.R. & Chinthamreddy, S.: Enhanced electrokinetic remediation of heavy metals in glacial till soils using different electrolyte solutions. *J. Environ. Eng.* 130 (2004), pp. 442–455.

Reddy, K.R., Chinthamreddy, S. & Al-Hamdan, A.Z.: Synergistic effects of multiple metal contaminants on electrokinetic remediation of soils. *J. Environ. Cleanup Costs Technol. Techniques* 11 (2001a), pp. 85–109.

Reddy, K.R., Xu, C.Y. & Chinthamreddy, S.: Assessment of electrokinetic removal of heavy metals from soils by sequential extraction analysis. *J. Hazard. Mater.* B84 (2001b), pp. 279–296.

Reddy, K.R., Danda, S. & Saichek, R.E.: Complicating factors of using ethylenediamine tetraacetic acid to enhance electrokinetic remediation of multiple heavy metals in clayey soils. *J. Environ. Eng.* 130 (2004), pp. 1357–1366.

Reuss, F.F.: Sur un novel effet de l'electricite galvanique. *Memoires de la Societe Imperiale des Naturalistes de Moscou* 2 (1809), pp. 327.

Ribeiro, A.B., Mateus, E.P., Ottosen, L.M. & Bech-Nielsen, G.: Electrodialytic removal of Cu, Cr and As from chromate copper arsenate-treated timber waste. *Environ. Sci. Technol.* 34(2000), pp. 784–788.

Ryu, B.G., Yang, J.S., Kim, D.H. & Baek, K: Pulsed electrokinetic removal of Cd and Zn from fine-grained soil. *J. Appl. Electrochem.* 40 (2010), pp. 1039–1047.

Shackelford, C.D.: *Bulletin of Transportation Research 1219*. Transportation Research Board, National Research Councuil, Washington DC, 1990, p. 23.

Shapiro, A.P.: *Electroosmotic purging of contaminants from saturated soils*. PhD Thesis, Department of Mechanical Engineering, Massachusetts Institute of Technol. MA, 1990.

Sivapullaiah, P.V. & Nagendra Prakash, B.S.: Electroosmotic flow behavior of metal contaminated expansive soil. *J. Hazard. Mater.* 143 (2007), pp. 682–689.

Suèr, P., Gitye, K. & Allard, B.: Speciation and transport of heavy metals and macroelements during electroremediation. *Environ. Sci. Technol.* 37 (2003), pp. 177–181.

Traina, G., Morselli, L. & Adorno, G.P.: Electrokinetic remediation of bottom ash from municipal solid waste incinerator. *Electrochim. Acta* 52 (2007), pp. 3380–3385.

Turer, D. & Genc, A.: Assessing effect of electrode configuration on the efficiency of electrokinetic remediation by sequential extraction analysis. *J. Hazard. Mater.* B119 (2005), pp. 167–174.

USAEC: In situ electrokinetic remediation of metal contaminated soils. Technology Status Report, US Army Environmental Center, Report Number SFIM-AEC-ET-CR-99022, USA, 2000.

USEPA: In situ electrokinetic extraction system. SITE technology capsule. United States Environmental Protection Agency, Office of Research and Development, Report Number EPA/540/R-97/509a, Cincinati, OH, 1998.

USEPA: Recent developments for in situ treatment of metal contaminated soils. Office of Solid Waste and Emergency Response, Technology Innovation Office, Washington, DC, 2005.

USEPA: Electrochemical design associates, lead recovery technology evaluation. United States Environmental Protection Agency, Office of Research and Development, National Risk Management Research Laboratory, Report Number EPA/540/R-04/506, Cincinati, OH, 2003.

Vengris, T., Binkiené, R. & Sveikauskaité, A.: Electrokinetic remediation of lead-, zinc- and cadmium-contaminated soil. *J. Chem. Technol. Biotechnol.* 76 (2001), pp. 1165–1170.

Virkuytyte, J., Sillanpää, M. & Latostenmaa, P: Electrokinetic soil remediation-critical overview. *Sci. Total Environ.* 289 (2002), pp. 97–121.

Wang, J.Y., Zhang, D.S., Stabnikova, O. & Tay, J.H.: Evaluation of electrokinetic removal of heavy metals from sewage sludge. *J. Hazard. Mater.* B124 (2005), pp. 139–146.

Wittle, J.K., Pamukcu, S., Bowman, D., Zanko, L.M. & Doering, F.: Field studies on sediment remediation. In: K.R. Reddy & C. Cameselle (eds): *Electrochemical remediation technologies for polluted soils, sediments and groundwater.* John Wiley & Sons, Inc., New Jersey, pp. 3–28.

Yang, J.W. & Lee, Y.J.: Electrokinetic removal of PAHs. In: K.R. Reddy & C. Camesselle (eds): *Electrochemical remediation technologies for polluted soils, sediments and groundwater.* Science Publishers, Inc., NH, 2009, pp. 197–217.

Yeung, A.T. & Gu, Y.Y.: A review on techniques to enhace electrochemical remediation of contaminated soils. *J. Hazard. Mater.* 195 (2011), pp. 11–29.

Yuan, C. & Weng, C.H.: Electrokinetic enhancement removal of heavy metals from industrial wastewater sludge. *Chemosphere* 65 (2006), pp. 88–96.

Zhang, P., Jin, C., Zhao, Z. & Tian, G.: 2D crossed electric field for electrokinetic remediation of chromium contaminated soil. *J. Hazard. Mater.* 177 (2010), pp. 1126–1133.

Zhou, D.M., Deng, C.F., Cang, L. & Alshawabkeh, A.N.: Electrokinetic remediation of a Cu–Zn contaminated red soil by controlling the voltage and conditioning catholyte pH. *Chemosphere* 61 (2005), pp. 519–527.

CHAPTER 6

Microbial *in-situ* mitigation of arsenic contamination in plants and soils

Nandita Singh, Pankaj Kumar Srivastava, Rudra Deo Tripathi, Shubhi Srivastava & Aradhana Vaish

6.1 BASICS OF ARSENIC BIOREMEDIATION

Arsenic (As) is present in the environment and poses threat to humans worldwide, mostly through food, water, and air. Excessive use of As-based pesticides and indiscriminate disposal of domestic (biosolids) and industrial (timber, tannery, paints, electroplating, etc.) wastes, as well as mining activities, have resulted in widespread anthropogenic As contamination of soils and water (Table 6.1). However, the presence of geogenic As in groundwater, a main source of drinking and irrigation water in many countries primarily in South Asia, has drawn much attention of the scientific community. Worldwide, more than 100 million people are exposed to excessive amounts of As in water. Arsenic is thus a geogenic contaminant-driving from natural sources, which is dissolved in groundwater and surface water through redox dissolution.

As causes multiple negative effects on human health. The US Environmental Protection Agency (USEPA) has classified inorganic As (i-As) as a known human carcinogen. Chronic exposure to As can cause cancer (Eguchi *et al.*, 1997). More than 70 million people are affected by As in India and Bangladesh alone (WHO, 2008). As contamination in the groundwater of West Bengal, India was first reported in the late 1980s (Bhattacharya *et al.*, 1997; Chakraborti *et al.*, 2004). As contamination has also been reported in groundwater from other states in India. Scores of people from India (Chakraborti *et al.*, 2002; Chatterjee *et al.*, 1995), Bangladesh (Smith *et al.*, 2000), China (Wang and Lazarides, 1984), Vietnam (Berg *et al.*, 2001), Taiwan (Lu, 1990), Chile (Smith *et al.*, 1998), Argentina (Hopenhayn-Rich *et al.*, 1998), and Mexico (Del Razo *et al.*, 1990) are

Table 6.1. Sources of arsenic in soils and aquatic environments.

Source	Concentration [mg kg^{-1}]	Reference
Coal	180	Wood (1996)
	2–825	Adriano *et al.* (1980)
Ores	2000	Tempel *et al.* (1977)
	500–9300	Hutchinson *et al.* (1982)
Fly ash	2–6300	Page *et al.* (1979)
	7000	Roussel *et al.* (2000)
Poultry manure	91.8	Abedin *et al.* (2002)
Rice straw	11.9–21.0	Ross *et al.* (1991)
Sewage sludge	2000	Walsh (1977)
Lead arsenate and other arsenical pesticides	100	Davenport and Peryea (1991)
	240	Aurelius (1988)
Waste disposal	3–350	Stilwell and Gorny (1997)
Wood preservative (chromated copper arsenate)	550	Cooper and Ung (1997)

likely at risk as well. As, a member of group V, occurs in nature in four states ($+5$, $+3$, 0 and -3), but the pentavalent arsenate [As(V)] and trivalent arsenite [As(III)] are the most common forms. Generally, iAs compounds are more toxic than organic As (org-As) compounds, and As(III) is more broadly toxic than As(V) (NRC, 1999).

Arsenic contaminated groundwater is not just used for drinking but is also widely used for irrigation of crops, and particularly for the staple food paddy/rice. The As content of lowland or paddy/rice grain is generally much higher than that of upland cereal crops (Schoof *et al.*, 1999; Williams *et al.*, 2007), because of the relatively high bioavailability of soil As under reduced conditions. A global range of 0.08 to 0.2 mg As kg^{-1} has been suggested for rice (Zavala *et al.*, 2008), but values as high as 1.8 mg As kg^{-1} have been found in Bangladesh rice (Meharg and Rahman, 2003). Compared to other countries, rice from Bangladesh and India had the highest percentage of i-As 80%, compared to 42% in rice from the USA. This indicates that the percentage of i-As in rice is not a constant factor geographically and probably depends on cultivar and growing conditions (Williams *et al.*, 2005).

Several studies have estimated As in non-rice foods (Alam *et al.*, 2003; Das *et al.*, 2004; Roy Chowdhury *et al.*, 2002; Williams *et al.*, 2006). They reported that As concentrations in vegetables, fruits, spices and freshwater fish, from As contaminated areas of Bangladesh and West Bengal, range from <0.04 to 3.99 mg kg^{-1} (dw). However, data from Taiwan showed that fish contained high level of i-As (Huang *et al.*, 2003). It has become clear that dietary exposure can contribute significantly to the total daily intake of i-As.

Several strategies exist to treat contaminated soils and water either through *ex-situ* or *in-situ* technologies. The use of biological system has recently gained importance over conventional methods. Bioremediation uses microorganisms to reduce, oxidize or eliminate contaminants. Its biological processes rely on biochemical changes induced by microbes. The common processes that are involved in bioremediation are chelation, compartmentalization, exclusion, sorption, biomethylation, complexation, co-precipitation, transformation, uptake and immobilization of different As species (Di *et al.*, 1999). In this chapter, the environmental and toxicological effects of As are discussed and followed by an overview of bioremediation techniques that can be employed for this element. Recent applications of As bioremediation are described in the chapter (Fig. 6.1).

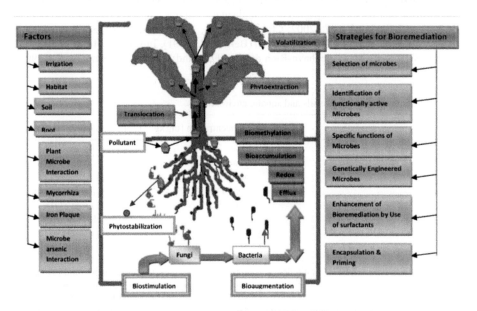

Figure 6.1.   Diagrammatic representation of *in-situ* mitigation of arsenic pollutant by plants and microbes including affecting factors and strategies for bioremediation.

The strategies for mitigation of As contamination related to agricultural activity along with scope of its future application are also highlighted.

6.2 INFLUENCE OF MICROBES ON THE SPECIATION AND BIOAVAILABILITY OF ARSENIC

6.2.1 *Arsenic speciation*

As is normally not associated with life, however it has been observed that various microorganisms gain energy for their growth from this element (Oremland and Stolz, 2003). These organisms play an important role in As speciation (Tables 6.2 and 6.3). For example, aqueous As in the $+3$ oxidation state, As(III), can be oxidized to As(V) by chemoautotrophic arsenite-oxidizing bacteria (Oremland and Stolz, 2005). There are also heterotrophic arsenite oxidizers that need organic carbon as energy source. Microbes use As(V) as an electron acceptor in anaerobic respiration, producing of As(III). i-As species can also be methylated to monomethyl As (MMA), dimethyl As (DMA) and trimethyl arsine (TMA) oxide (Cullen and Reimer, 1989). DMA can be transformed by microorganisms via two pathways: (i) reductive conversion to volatile organo-arsine species (e.g., dimethyl- or trimethyl arsine) and emissions from the soil system; and (ii) demethylation to produce the end products CO_2 and As(V); the first pathway predominates under anaerobic conditions, whereas both pathways occur in aerobic soil (Woolson and Kearney, 1973; Yoshinaga *et al.*, 2011). The rate of detoxification and the relative importance of the two pathways vary among different studies, probably due to impact of different soil properties (pH, water logging, site hydrology, behavior of soil colloids), microbial communities and environmental conditions (Sadiq, 1997).

Table 6.2.   Bacterial classes investigated for environmental remediation of arsenic contamination.

Bacterial class	Mechanism	Reference
Gammaproteobacteria	Oxidizes arsenite to arsenate	Butt and Rehman (2011), Nagvenkar and Ramaiah (2010), Srivastava *et al.* (2010), Chitpirom *et al.* (2009), Aksornchu *et al.* (2008), Saltikov and Olson (2002)
Actinobacteria	Removal and transformation	Nagvenkar and Ramaiah (2010)
Proteobacteria	Oxidizes arsenite to arsenate, Arsenic resistant	Chitpirom *et al.* (2009), Macur *et al.* (2004), Caudill (2003)
Bacilli	Biosorption, Reduces arsenate to arsenite, Biomethylation	Aksornchu *et al.* (2008), Yamamura *et al.* (2007), Jenkins *et al.* (2003)
Betaproteobacteria	Biosorption, reduces arsenate to arsenite	Aksornchu *et al.* (2008), Santini and Vanden Hoven (2004)
Aeromonasbacteria	Arsenic resistant	Pepi *et al.* (2007)
Corynebacteria	Arsenic tolerant	Chang *et al.* (2008)
Flavobacteria	Arsenic resistant	Macur *et al.* (2004)
Alphaproteobacteria	Reduces arsenate to arsenite	Santini and Vanden Hoven (2004), Macur *et al.* (2001)
Deltaproteobacteria	Reduces arsenate to arsenite	Lloyd and Oremland (2006), Michalke *et al.* (2000), Macy *et al.* (2000)
Methanobacteria, Clostridia	As methylation and demethylation	Michalke *et al.* (2000)
Thermos	Reduces arsenate to arsenite, Oxidizes arsenite to arsenate	Gihring and Banfield (2001), Gihring *et al.* (2001)
Desulfitobacterium	Reduces arsenate to arsenite	Niggemeyer *et al.* (2001)

Table 6.3.   Fungi investigated for environmental remediation of arsenic contamination.

Fungi	Mechanism	Reference
Ascomycota	Tolerance, accumulation, biosorption, biovolatilization, and removal	Srivastava *et al.* (2011), Su *et al.* (2010), Wysocki and Tamas (2010), Adeyemi (2009), Cernansky *et al.* (2009), Maheswari and Murugesan (2009), Vala *et al.* (2010), Cernansky *et al.* (2007), Buckova *et al.* (2007), Murugesan *et al.* (2006), Pokhrel and Viraragahavan (2006), Canovas *et al.* (2003), Lehr *et al.* (2003), Granchinho *et al.* (2002), Sharples *et al.* (2000), Hofman *et al.* (2001), Visoottiviseth and Panviroj (2001), Mukhopadhyay *et al.* (2000)
Basidiomycota	Bioaccumulation	Adeyemi (2009), Soeroes *et al.* (2005), Demirbas (2001), Hofman *et al.* (2001), Lehr *et al.* (2003)
Glomeromycota	Tolerance	Xu *et al.* (2008)
Zygomycota	Biosorption, tolerance and bioaccumulation	Srivastava *et al.* (2011), Bai and Abraham (2003)

6.2.2   *Role of soil*

The toxicological effects of As depend upon its chemical form and bioavailability (La Force *et al.*, 2000). The hydrated forms are considered to be the most toxic forms of As, and strong complexes and species associated with colloidal particles are usually assumed to be less toxic in soils (Russeva, 1995). The toxicity of As depends on various soil properties, for e.g., water saturation/logging, pH redox conditions, other elements (phosphorus, silica and selenium), and site hydrology. Plant and microbial components influence the adsorption capacity and behavior of soil colloids (clay, metal oxides or hydroxides, calcium carbonate and/or organic matter) and these effects may regulate solubility and bioavailability of As (Sadiq, 1997). In general, iron oxides/hydroxides are most commonly involved in the adsorption of As in both acidic and alkaline soils (Polemio *et al.*, 1982). As is known to adsorb to Fe/Mn oxyhydroxides, clays, carbonate and organic matter (Dixit and Hering, 2003; Goldberg, 2002; Ongley *et al.*, 2007; Romero *et al.*, 2004). In soils contaminated by mining activities, As is primarily associated with amorphous iron oxyhydroxides in soils (Ahumada *et al.*, 2004; Filippi *et al.*, 2004; Ghosh *et al.*, 2004). As can also form secondary minerals, such as scorodite and sulfide minerals, or can co-precipitate with other minerals (Fendorf *et al.*, 2004; Filippi *et al.*, 2004).

6.2.3   *Role of microbes*

As stated earlier, microorganisms play an important role in the environmental fate of As with a multiplicity of mechanisms affecting transformations between soluble and insoluble As forms and toxic and nontoxic As forms (Turpeinen *et al.*, 2002). Bacteria *Pseudomonas arsenitoxidans* can derive metabolic energy from As(III) oxidation (Ilyaletdinov and Abrashitova, 1981; Anderson *et al.*, 2003), on the other hand, As(V) can be reduced by dissimilatory reduction where microbes utilize As(V) as a terminal electron acceptor for anaerobic respiration. This has been observed in several bacterial species including *Sulfurospirillum barnesii, S. arsenophilum, Desulfotomaculum auripigmentum, Bacillus arsenicoselenatis, B. selenitireducens, Crysigenes arsenatis, Sphingomonas* sp., *Pseudomonas* sp. and *Wolinella* sp. (Ahmann *et al.*, 1994; Lovley and Coates, 1997; Macur *et al.*, 2001; Newman *et al.*, 1997; 1998; Stolz and Oremland, 1999; Oremland *et al.*, 2000).

Microbes can impact As mobility through indirect natural processes such as oxidative sulfide mineral dissolution, reduction of iron oxides, and sulfate reduction. Direct microbial processes such as As(V) reduction and As(III) oxidation can impact As abundance and speciation as well. Arsenite oxidation by microorganisms is potentially significant process, which depends upon bio-catalytic activities of microorganisms (Gihring and Banfield, 2001; Gihring *et al.*, 2001). Some microbial strains possess genetic determinants that confer resistance to As toxicity (Turpeinen *et al.*, 2000; Weeger *et al.*, 1999). Many bacteria are known for their ability to transform i-As species by redox reaction (Simeonova *et al.*, 2004). The bacterial oxidation of As(III) to As(V) represents a detoxification mechanism with its potential scope in bioremediation because it generates less toxic and less mobile forms of As (Oremland, 2002).

Speciation of As has been studied using various microbes, (*Staphylococcus aureus, Bacillus subtilis* and *Escherichia coli*) (Tauriainen *et al.*, 1997; Turpeinen *et al.*, 1999; 2002). It is known that As(V) is subjected to microbial reduction and methylation leading to volatilization as arsines (Alexander, 1977; Gao and Barau, 1997). However, the reduction and/or methylation rates of As, which are necessary pre-requisites for production of arsine, vary greatly depending on the properties of the matrix, such as temperature, different species of As, and microbial populations. Macy *et al.* (2000) reported *Desulfomicrobium* sp. BenR-B showing reduction of As(V) to As(III) via enzyme arsenate reductase. The mechanisms involved in the microbial transformation and removal of As from the contaminated matrix included adsorption via reduction reaction (*Desulfomicrobium* sp. BenR-B), oxidation/reduction reaction (*Trichoderma harzianum* AS11 and *Trichosporon mucoides* SBUG801), and methylation reaction (*Paenibacillus* sp. and *Pseudomonas* sp.) (Macy *et al.*, 2000). Fungi methylate As for detoxification and are producing MMA or DMA. Granchinho *et al.* (2002) reported that the fungus *Fusarium oxysporum melonis*, isolated from the alga *Fucus gardneri*, is capable of accumulating As(V) from the surrounding medium and transforming it into As(III) and further methylated species. The predominant As species found in the MeOH-H$_2$O extracts of the fungus after As(V) exposure were As(III), DMA species at concentrations about three times less than found in the exposed fungus, i.e., 500 ppb.

Both the Archaeal and Bacterial Domains can also produce volatile methylated arsines, which remove As from contaminated matrix by forming a gaseous compound. An alternative strategy used by bacteria and yeast (*Saccharomyces cerevisiae*) is based upon arsenate reductase "arsC operon/genes". Earlier, Challenger (1945) revealed that trimethylarsine was produced by *Scopulariopsis brevicaulus*, from As(III). Subsequent work confirmed that other fungi such as *Penicillium* sp., *Gliocladium roseum* and the yeast *Candida humicola* were also capable of As biomethylation (Andreae, 1986; Challenger, 1978; Thayer, 1984). Bacteria, including *Aeromonas* sp., *Flavobacterium* sp. and *Escherichia coli* can produce a variety of methylated As derivatives including dimethyl- and trimethylarsine. Turpeinen *et al.* (2004) observed a positive relationship between the bioavailability of As and the proportion of As(III) in As resistant species of *Acinetobacter, Edwardsiella, Enterobacter, Pseudomonas, Salmonella* and *Serratia*, which are classified into Betaproteobacteria or Gammaproteobacteria.

Speciation of As in soil has become an important issue due to the different toxicity levels of i-As(V) and As(III) species and gaseous species such as arsine, MMA, DMA and TMA. Because microbial processes play a major role in controlling the speciation of As in soil, it is important to study how microbial activities affect As biogeochemistry. It is important to understand microbially-mediated mechanisms involved in bioremediation of As-contaminated soils and associated risks.

6.3 MITIGATION OF AS CONTAMINATION IN SOIL: MICROBIAL APPROACHES AND MECHANISMS

Bioremediation of As contaminated soils and water shows a great potential for cost effective environmental improvement. Several microbes-mediated processes like reduction, mobilization, immobilization through sorption, biomethylation, complexation, and oxidation-reduction can be used for bioremediation of different As species in soils. These processes basically involve organic carbon, Fe, Mn and S, affecting As mobility. As stated earlier, microorganisms have also evolved biochemical mechanisms to exploit As oxyanions, either as an electron acceptor [As(V)] for anaerobic respiration, or as an electron donor [As(III)] to support chemoautotrophic fixation of carbon dioxide into cell carbon (Santini *et al.*, 2000; Silver and Phung, 2005; Stolz and Oremland, 1999; Wang and Zhao, 2009; Zobrist *et al.*, 2000). A number of methodologies have been investigated for bioremediating As-contaminated soils.

6.3.1   *Biostimulation*

Biostimulation include the introduction of adequate amounts of water, nutrients and oxygen into the soil to enhance the activity of bioremediating indigenous microbial agents (Couto *et al.*, 2010) and/or to promote co-metabolism (Lorenzo, 2008). The concept of biostimulation is to boost the intrinsic production of enzymes participating in bioremediation process of a polluted matrix and has been used for a wide variety of xenobiotics (Abdulsalam *et al.*, 2011; Kadian *et al.*, 2008).

Even though the diversity of natural microbial populations apparently displays the potential for contaminant remediation at polluted sites, the factors such as lack of electron acceptors or donors, low nitrogen or phosphorus availability, or a lack of stimulation of the metabolic pathways are responsible for inhibition or delay of the bioremediation process. In these cases, accumulation of exogenous nutrients can enhance the remediation of the toxic materials (Cosgrove *et al.*, 2010). Elemental sulfur can be added as an energy substrate and an acid source in aerobic conditions to stimulate As leaching from soil and sediments by microbes. Seidel *et al.* (2002) studied the bioleaching of As from a highly polluted lake sediment (753 mg As kg^{-1}) by indigenous *Thiobacillus* sp. under aerobic and anaerobic conditions. Under aerobic conditions without adding elemental sulfur, the As solubility ranged between 0.6–3.5 mg kg^{-1}. However, stimulating the aerobic bioleaching with elemental sulfur, the total soluble As was increased up to 80% (660 mg kg^{-1}) in the form of As(III) and As(V) species. In the presence of sulfur, soluble As gets immobilized. Kohler *et al.* (2001) reported that release of As ions and soluble organo As compounds from contaminated soil by autochthonic soil bacteria and a mixture of isolated pure cultures were increased after addition of nitrogen, phosphorus, sodium acetate and ethanol. Chatain *et al.* (2005) found an increase of As bioleaching by 28-folds by indigenous anaerobic bacteria by addition of carbon sources and this may help in mitigating As contamination in plants.

Biostimulation technique has already been extended to the removal of wide array of environmental contaminants. The processes of biostimulation can be successfully applied for remediation of As-contaminated soil and water.

6.3.2   *Bioaugmentation*

Addition of previously grown microbial cultures to enhance the remediation of contaminants is called bioaugmentation. Bioaugmentation, an economical and eco-friendly approach, has emerged as the most advantageous soil and water clean-up technique for contaminated sites containing heavy metals and/or organic pollutants (Tyagi *et al.*, 2011). It is done in conjunction with the development and monitoring of an ideal growth environment, in which selected microbes can live and work. Bioaugmentation with microbes plays an important role in biogeochemical cycling of As and in As detoxification by various processes, such as oxidation-reduction, efflux, biosorption, bioaccumulation and biomethylation.

6.3.2.1   *Microbially mediated As(V) reduction and As(III) oxidation*
The understanding of biogeochemical processes is crucial for predicting and protecting environmental health and can provide new opportunities for remediation strategies. Energy can be released and stored by means of redox reactions via the oxidation of labile organic carbon or inorganic compounds (electron donors) by microorganisms coupled with the reduction of electron acceptors including humic substances, iron-bearing minerals, transition metals, metalloids, and actinides. Environmental redox processes play key roles in the formation and dissolution

of mineral phases. Redox cycling of naturally occurring trace elements and their host minerals often controls the release or sequestration of inorganic contaminants. Redox processes control the chemical speciation, bioavailability, toxicity, and mobility of many major and trace elements (Borch *et al.*, 2010). Anaerobic microorganisms can potentially use As(V) as an electron acceptor for the oxidation of organic matter, yielding energy to support their growth (McLaren and Kim, 1995; Oremland and Stolz, 2003). Microorganisms, including bacteria, archaea and fungi, display resistance to As(V) toxicity, which involves a common mechanism of resistance by the reduction of intracellular As(V) to As(III) by As(V) reductase with As(III) being pumped out via efflux pumps (Oremland and Stolz, 2003).

Microbial reduction of As(V) to As(III) has been proposed to contribute to As mobilization (Tufano *et al.*, 2008a), although this may not be universally true (Campbell *et al.*, 2008). Several biogeochemical processes can directly or indirectly lead to redox transformations of As. The reduction of As(V) to As(III) under anaerobic conditions has been reported to be mediated by a diverse populations of anaerobic microorganisms, including methanogens, fermentive bacteria, and sulfate- and iron- reducers. This indicates that arsenate can act as terminal electron acceptor for anaerobic respiration (dissimilatory arsenate reduction).

Known As(III) oxidizing bacterial strains are distributed in more than 20 genera and have been isolated from various environments. They include both chemolithotrophs such as *Acidicaldus* sp. and *Acidithiobacillus* sp. (Brierley and Brierley, 2001; D'Imperio *et al.*, 2007) and heterotrophs such as *Agrobacterium* sp., *Alcaligenes* sp., *Burkholderia* sp., *Thiomonas* sp., *Acinetobacter* sp. and *Pseudomonas* sp. (Cai *et al.*, 2009; Krumova *et al.*, 2008; Quéméneur *et al.*, 2008; Santini and Vanden Hoven, 2004). As resistant bacteria are usually associated with its ability to reduce As(V) to As(III) (Huang *et al.*, 2010). Study of *aox* gene transcripts (Quéméneur *et al.*, 2008) or the *Ars* and *Acr* transporter genes transcripts would help for better understanding of the processes involved in As(III) oxidation.

Microorganisms may further indirectly induce As(III) oxidation or As(V) reduction. In particular, they can produce reactive organic or inorganic compounds that subsequently undergo redox reactions with As(V) or As(III). According to Messens *et al.* (2002), *Staphylococcus aureus* pI258and *Bacillus subtilis* are expressing a thioredoxin-coupled arsenate reductase (ArsC). The ArsC from *E. coli* plasmid R773 and ACR1, ACR2 and ACR3 were identified on chromosome XVI of *Saccharomyces cerevisiae*. ACR1 encodes a transcription regulatory protein. Eukaryotic ACR2p encodes the arsenate reductase. ACR3 encodes the ACR3p membrane transporter that effluxes arsenite from the cells. ACR2p from *Saccharomyces cerevisiae* represent two distinct glutaredoxin-linked ArsC classes. All are small cytoplasmic redox enzymes that reduce arsenate to arsenite by the sequential involvement of three different thiolate nucleophiles that function as a redox cascade (Bobrowicz *et al.*, 1997; Ordonez *et al.*, 2009). In case of fungi, *Aspergillus* sp. strain P37, an arsenate hyper-tolerant strain, isolated from the As polluted Rio Tinto in South Western Spain showed that arsenate triggered an increase in the accumulation of GSH. Such an accumulation could contribute to arsenate detoxification as GSH can serve as the electron donor in enzyme-catalyzed arsenate reduction and GSH binds arsenite to form As(GS)$_3$, which allows sequestration in vacuoles mediated by an ATP binding cassette-type glutathione conjugate transporters (Rosen, 2002). This indicates that *Aspergillus* sp. strain P37, which maintained higher GSH levels and has higher arsenite efflux following arsenate exposure in comparison to sensitive *A. nidulans* TS1, has enhanced arsenate reductase capacity (Cánovas, 2004).

6.3.2.2 *Bioaccumulation and biosorption*

Bioaccumulation of As mainly involves the biosorption of As by microbial biomass and its byproducts; and physiological uptake of As by microorganisms through metabolic processes. Microorganisms can take up As(V) via phosphate transporters (Zhao *et al.*, 2009), and then reduce the As(V) internally to As(III), which is then either extruded from the cells or sequestered in intracellular components (either as free As(III) or as conjugates with glutathione or other thiols) (Mateos *et al.*, 2006). Takeuchi *et al.* (2007) reported that an As resistant isolate *Marinomonas communis* was capable of removing 2290 mg kg^{-1} As from culture medium amended with 5 mg As(V) L^{-1} by accumulation in their cells. As(III) in water is an inorganic equivalent of non-ionized glycerols and can be transported across cell membranes by glyceroporin channel proteins (Rosen, 2002; Ma *et al.*, 2008).

The intracellular uptake of metal ions from a substrate into living cells may lead to the biological removal of metals by fungi. Su *et al.* (2010) reported that fungi (*Penicillium janthinellum*, *Fusarium oxysporum* and *Trichoderma asperellum*) bioaccumulate As(V) under laboratory conditions. *Trichoderma asperellum* and *Fusarium oxysporum* showed superior abilities for the absorption of extracellular As and accumulation of intracellular As, which accounted for 82.2 and 63.4% of the total accumulated arsenic, respectively. In contrast, *Penicillium janthinellum* presented an equal distribution of intracellular and extracellular As. According to Adeyemi (2009), three filamentous fungi (*Aspergillus niger*, *Serpula himantioides* and *Tremetes versicolor*) were investigated for their potential abilities to bioaccumulate As. Accumulation of As in the fungal biomass was observed in the order of *Tremetes versicolor* > *Serpula himantioides* > *Aspergillus niger*.

Biosorption is the passive sequestration of metals and metalloids by live or dead biological mass, which is at present the most practical and widely used approach for the bioremediation of metals, and metalloids. Biosorption is effective in treating water and wastewater (Schiewer and Volesky, 2000), but its potential in soil is less attempted (Ledin, 2000). Ion exchange, adsorption, microprecipitation, and electrostatic and hydrophobic interactions facilitate biosorption. Whereas mechanisms of metal binding by individual cellular organelles and chemical moieties are known (Schiewer and Volesky, 2000), sorption of metals to intact cells or microbial biomass is governed by a multiplicity of mechanisms and interactions and thus not always fully understood. Langley and Beveridge (1999) attempted to understand the role of carboxyls in the binding of metal cations to *O*-side chains of lipopolysaccharide (LPS) and concluded that metals bound most likely to phosphoryl groups in the LPS. This negatively charged side chains influence binding of metals to Gram-negative bacteria by affecting cell hydrophobicity. The kinetics of metal binding onto fungal biomass depends upon heterogeneous non-equivalent interactions, multiplicity of binding sites, charge types, accessibilities and the properties of bound metals (Gadd, 2000).

According to Volesky and Holan (1995), biosorption includes several mechanisms such as ion exchange, chelation, adsorption and diffusion through cell walls and membranes. These mechanisms may differ depending upon the fungal species used, the origin and processing of the biomass and solution chemistry (Ceribasi and Yetis, 2001). The metal biosorption abilities of different fungi (*Rhizopus nigricans* and *Mucor rouxii*) have been previously reported (Bai and Abraham, 2003; Yan and Viraraghavan, 2003). Biosorption of toxic metals is based upon ionic species associating with the fungal cell surface such as extracellular polysaccharide, proteins and chitins of fungi (Zafar *et al.*, 2007). Metal sorption activity of fungal cells depends on structural cell wall polysaccharides and chitin:glucan ratio (Tereshina *et al.*, 1999). Many fungi contain chitin and chitosan as integral parts of their cell wall structure and these are effective biosorbents for metals (Gadd, 2004). The deacetylated amino group of glucosamine of chitosan may act as binding sites for metals (Pillichshammer *et al.*, 1995). Zhou (1999) also revealed that in fungus *Rhizopus arrhizus* metal biosorption occurred due to wall chitin and chitosan. The biomass of the fungus *Penicillium purpurogenum* has been found effective in biosorption of As (Say *et al.*, 2003). The maximum adsorption capacities of heavy metal ions onto the fungal biomass under noncompetitive conditions were 35.6 mg g^{-1} for As(III), 70.4 mg g^{-1} for Hg(II), 110.4 mg g^{-1} for Cd(II) and 252.8 mg g^{-1} for Pb(II).

6.3.2.3   *Efflux*

Different biological systems have evolved diverse strategies to tolerate the toxic effect of As. Presumably, microorganisms bear at least one arsenate reductase and a membrane bound pump complex for the efflux of As(III) back to the external medium. It is perhaps an indication of the early evolutionary success of such a mechanism during the rise of an oxidizing atmosphere (Canovas *et al.*, 2004).

In microbes, a chromosomally encoded efflux system transports As using either a single-polypeptide system (*ArsB*) or a two-component system (*ArsA* and *ArsB*), which functions as a chemiosmotic transporter (Lee *et al.*, 2001; Silver *et al.*, 1982). The *ArsB* protein is a membrane protein that functions alone as a chemiosmotic As(III) transport protein (Silver, 1996), whereas the *ArsA* gene encodes a unique ATPase that binds to *ArsB* (Wu *et al.*, 1992). The *ArsAB* pump is composed of six transmembrane segments and a catalytic subunit that functions in the reduction mechanism of the *ArsA* ATPase (Bruhn *et al.*, 1996). Efflux mechanism has been reported in *Pseudomonas putida* and *Escherichia coli* (Chang *et al.*, 2007).

Mycorrhizal fungus *Hymenoscyphus ericae* demonstrated an enhanced As efflux mechanism in comparison with nonresistant *Hymenoscyphus ericae* and lost approximately 90% of preloaded cellular As(III) (Sharpels *et al.*, 2000). The archetypical yeast *Saccharomyces cerevisiae* lacks functional phytochelatins, and its As tolerance depends upon the efflux of As across the plasma membrane and on the vacuolar sequestration of As(GS)$_3$ (Ghosh *et al.*, 1999).

6.3.2.4 *Biomethylation and biovolatilization*

As undergoes biomethylation to form non-volatile monomethylarsonic acid (MMAA), dimethylarsenic acid (DMAA), volatile trimethylarsine oxide (TMAO) and trimethylarsine [TMA(III)] (Bentley and Chasteen, 2002; Zeng *et al.*, 2010). The expression of As methyl transferase (*ArsM*) in bacteria *Rhodopseudomonas palustris* has been found to increase As tolerance in *E. coli* strain AW 3110, in which the *Ars RBC* operon was detected (Qin *et al.*, 2006). This *ArsM* was the first arsenite-*S*-adenosylmethyltransferase identified in bacteria. It mediates bacterial As resistance by catalyzing the formation of dimethylarsenate [DMAs(V)], trimethyl arsine oxide [TMAs(V)O] and trimethylarsine [TMAs(III)] gas that can leave the cell due to its volatility (Cullen and Bentley, 2005; Qin *et al.*, 2006; Yuan *et al.*, 2008). Meng *et al.* (2011) constructed transgenic rice with the *ArsM* gene from *R. palustris* and demonstrated that the resulting transgenic rice plant acquired the capability of volatilization of As. This is the first report of *in planta* methylation and volatilization. Two other As methyl transferase gene (termed *CmarsM7* and *CmarsM8*) were cloned and characterized from the unicellular eukaryotic red alga *Cyanidroschyzon* from the Yellowstone National Park, USA (Qin *et al.*, 2009).

The conversion of As(V) to volatile methylarsines was described in a pure culture of a methanogen *Methanobacterium bryantii* (McBride and Wolfe, 1971). Recently, several pure cultures of anaerobes, including a methanogen (*M. formicicum*), a fermentative bacterium (*Clostridium collagenovorans*) and sulfate-reducing bacteria (*Desulfovibrio vulgaris* and *D. gigas*), were also capable to form methylarsines (Michalke *et al.*, 2000). As(V) can be converted to monomethylarsine and dimmethylarsine by *Achromobacter* sp. and *Enterobacter* sp., and to monomethylarsine, dimethylarsine and trimethylarsine by *Aeromonas* sp. and *Nocardia* sp. (Cullen and Reimer, 1989).

Fungi are also able to transform inorganic and org-As compounds into volatile methylarsines (Tamaki and Franekanberger, 1992). Some yeast and other fungi such as *Candida humicola*, *Glicladium roseum* and *Penicillium* sp., are capable of converting MMA and DMA to TMAO (Cox and Alexander, 1973). Effective biovolatalization ($\sim$23% of As was volatilized from all culture media) of As was observed in the heat–resistant *Neosartorya fischeri* strain, while transformation of As to volatile derivates was approximately two times lower than the non-heat-resistant *Aspergillus niger* strain (Cernansky *et al.*, 2007). The order of ability of As biovolatalization was observed in the same study as *Neosartorya fischeri* > *A. clavatus* > *A. niger* (Cernansky *et al.*, 2009). As resistant fungi *Pennicillum janthinellum*, *Trichoderma asperellum* and *Fusarium oxysporum* also accumulate and volatilize As (ranging from 100 to 304.06 μg) from culture medium (Su *et al.*, 2010). As biovolatilization has also been reported in *Trichoderma* sp., *Rhizopus* sp., *Penicillum* sp., and *Aspergillus* sp. within a range of 3–29% (Srivastava *et al.*, 2011). The fungi-mediated biovolatilization process in agriculture soils may lead to reduction of As load in those contaminated soils.

6.4   MICROBES-MEDIATED MITIGATION OF AS IN CONTAMINATED SOILS: ASSOCIATED FACTORS AFFECTING MITIGATION

The mitigation of As by microbes will depend on the amount of available As, soil properties and other environmental conditions (Brammer, 2009). Following major factors are having roles in As contamination to crops.

6.4.1   *Irrigation*

The problem of As in plants and crops predominantly occurs in the areas where As-contaminated groundwater is being used to irrigate agriculture lands. When soils kept submerged as in flooded paddy fields, As gets mobilized into the soil solution mainly as arsenite (Xu *et al.*, 2008). This is a result of reduction of strongly adsorbed As(V) to more weakly adsorbed As(III), leading to an enhanced partition of As from the solid to the solution phase. Consequently the bioavailability of As is enhanced to rice plants grown under submerged conditions than to those grown under aerobic conditions. The As conversion in soil to different bioavailable forms is affected by water management. Soil bacteria are also mobilizing As into irrigation water.

6.4.2   *Habitat*

Variation exists in the environmental conditions of affected habitats such as climate, sediment sources, geomorphology and hydrology etc. The diverse environment of contaminated habitats regulates the microbiota and their activities directly through the modification of their population size, diversity and biogeochemistry under different toxicity levels (Kavamura and Esposito, 2010). Kandeler *et al.* (1996) noticed that an increase in the contaminant level of heavy metals above their threshold limits in the soil exerted a negative influence on microbial activities, resulting in a decrease in the functional diversity of soil microorganisms and activities of soil enzymes in those contaminated soils.

6.4.3   *Soil properties*

As chemistry in the soil is affected by soil chemical and biological properties. The survival, propagation and functioning of microbes in the rhizosphere noticeably vary for a range of soil processes (Hinsinger *et al.*, 2005; Watt *et al.*, 2006), These processes depend not only on the host plants (Kamilova *et al.*, 2006; Nunan *et al.*, 2005) but also on the conditions and properties of the medium, e.g., soil and water composition of habitat and concentrations of nutrients and pollutants, as well as of other competing microorganisms are majorly influential in regulating these processes (Gunatilaka, 2006; Kamilova *et al.*, 2006). Soil pH is one of the main factors controlling the solubility of heavy metals; and solubility increases when the pH falls below 6.0 (Ross, 1994). Addition of lime with organic matter may raise the soil pH of acidic soils (having pH <6.5) and immobilize the metals (Clement *et al.*, 2003). Further this also depends on the nature of the organic matter and microbial population creating the redox conditions for As speciation.

6.4.4   *Root exudates*

The number of microorganisms in the rhizosphere is typically one order of magnitude larger than in non-rhizosphere soil due to the continuous input of root derived organic substrates, resulting in a more diverse, active and synergetic microbial community (Marschner, 1995). The production of organic acids, such as oxalic acid by ectomycorrhizal fungi (Malajczuk and Cromack, 1982), could also be of importance in determining metal uptake, since these acids can bind metals.

Root mediated rhizospheric interactions could be positive or negative. Positive interactions involve root exudate mediated interactions with plant growth-promoting rhizobacteria (PGPR), mycorrhizae and rhizobia. Roots produce chemical signals that attract bacteria and induce chemo-taxis (Thimmaraju *et al.*, 2008). Positive interactions mediated by root exudates include population of growth facilitators that support growth of plants. Negative interactions mediated by root exudates involve secretion of antimicrobial compounds (Bais *et al.*, 2006). Excretion of organic acids and CO_2 production by roots may affect rhizosphere microorganisms and the buffering capacity of the soil (Marschner, 1995). PGPR may (i) synthesize siderophores, which can solubilize and sequester iron from the soil and provide it to plant (Rajkumar *et al.*, 2010), (ii) organic acids of root exudates solubilize minerals such as phosphorus, which becomes readily available. Availability of Fe may affect the iron-plaque formation by microbes in roots of rice plants and consequent As adsorption. Being chemical analogs, both As and P have comparable dissociation constants for their acids and solubility products for their salts, resulting in similar geochemical behavior of As and P in soil (Adriano, 2001). Hence, it is reasonable to assume that carboxylate exudation could play a role in the bioavailability of As in the rhizosphere and its further uptake by plants.

6.4.5 *Plant microbe interactions*

Interactions between plants and soil microbes are highly dynamic in nature and based on co-evolutionary pressures (Morgon *et al.*, 2005; Reinhart and Callaway, 2006). Consequently, it is not astonishing that rhizosphere microbial communities differ among plant species, and further genotypes within species along with different developmental stages of a particular plant (Batten *et al.*, 2006; Mougel *et al.*, 2006; Wei *et al.*, 2007) (Table 6.4). *Pteris vittata*, an As-hyperaccumultor, has generated interest in As-phytoremediation (Ma *et al.*, 2001). Applications of microbes for enhancing phytoremediation in *P. vittata* has been attempted with positive results like plant biomass increased by >53% and As uptake by >44% (Agely *et al.*, 2005; Huang *et al.*, 2010; Liu *et al.*, 2005; Vogel-Mikus *et al.*, 2006; Yang *et al.*, 2011). The phytoremediation of As can be influenced by the variation of rhizospheric microbes; and the interactions between microbes and plant roots, including microbial and plant root exudates (Tang *et al.*, 2001).

6.4.6 *Mycorrhiza*

Arbuscular mycorrhizal (AM) fungi have ubiquitous symbiotic associations in both natural and contaminated sites (Wang *et al.*, 2007). AM fungi may stimulate metal phytoextraction by essentially improving plant growth and consequent increase in total metal uptake (Agely *et al.*, 2005; Barua *et al.*, 2010; Leung *et al.*, 2006). According to Leung *et al.* (2006), the infectious percentage of mycorrhizas was 26.4, 30.3 and 40.6% upon 0, 50 and 100 mg kg^{-1} As treatment/exposure, respectively, compared to the control. The indigenous mycorrhizas enhanced As accumulation as 3.70, 58.3 and 88.1 mg kg^{-1}, respectively. Mycorrhizal fungi may alleviate metal toxicity to the host plant by acting as a barrier for metal uptake in some cases (Leyval *et al.*, 1997). AM fungi have been shown to not only increase the nutrient status of their host plant, but also to improve the plant's ability in tolerating toxic elements (Aggangan *et al.*, 1998). Zhang *et al.* (2005) noted that AM fungi contributed to the resistance of upland rice grown in soil contaminated with metals. Li *et al.* (2011) showed that rice/AMF combinations had beneficial effects on lowering grain As concentration, improving grain yield and grain P uptake. Variation in transfer and uptake of As and P reflect functional diversity in AM symbioses in different plants.

6.4.7 *Iron plaque*

Liu *et al.* (2004) have shown that As can be sequestered in iron plaque of root surface of plants, thus reducing As uptake in plant tissues. Iron plaque is a precipitate of reddish-brown Fe oxides and is formed on to the roots of paddy rice (Liu *et al.*, 2004). Iron plaque is formed by abiotic oxidation by iron oxidizing bacteria (Weiss *et al.*, 2003). Studies have demonstrated that the microbial reduction of As-bearing Fe(III) (hydro)oxides results in a dissolution of the solid

Table 6.4.   Bioaugmentation of microbes in soil with plants for the remediation of arsenic.

Augmented microbes	Plants	Reference
Bacteria		
Rhodococcus sp. TS1 *Delftia* sp. TS33, *Comamonas* sp. TS37, *Delftia* sp. TS41, *Streptomyces lividans* sp. PSQ22 *Pseudomonas* sp., *Comamonas* sp., *Stenotrophomonas* sp.	*Pteris vittata* L.	Patel *et al.* (2007), Yang *et al.* (2011), Ghosh *et al.* (2011)
Agrobacterium radiobacter	*Populus deltoids* W. Bartramex Marshall	Wang *et al.* (2011)
Sinorhizobium meliloti Sinorhizobium sp.	*Medicago sativa* L.	Panigrahi and Randhawa (2010), Pajuelo *et al.* (2008)
Alphaproteobacteria, Achroobacter, Brevundimonas, Microbacterium, Ochrobactrum, Ancyclobacter dichloromethanicum	*Cirsium arvense* L. Scop.	Cavalca *et al.* (2010)
Azotobacter chroococcum	*Jatropha curcas* L.	Kumar *et al.* (2008)
Bacillus sp., *Serratia* sp., *Ochrobactrum* sp., *Arthrobacter, Acaulospora, Gigaspora*	*Agrostis tenuis* Sibth.	Chopra *et al.* (2007)
Bradyrhizobium japonicum CB1809	*Glycine max* cv. Curringa	Reichman (2007)
Enterobacter cloacae CAL2	*Brassica napus* cv. Westar	Nie *et al.* (2002)
MYCORRHIZA		
Arbuscular mycorrhizal fungi	*Gmelina arborea* Roxb. *Pteris vittata* L. *Cynodon dactylon* L. Pers.	Barva *et al.* (2010) Leung *et al.* (2006) Agely *et al.* (2005)
Glomus mosseae	*Zea mays* L. *Pteris vittata* L. *Trifolium repens* L. *Lolium perenne* L. *Coreopsis drummondii* Torr. and A. Gray *Solanum lycopersicum* L.	Yu *et al.* (2009), Wang *et al.* (2007) Liu *et al.* (2009; 2005a), Chen *et al.* (2006; 2007), Wu *et al.* (2009), Trotta *et al.* (2006) Dong *et al.* (2008), Chen *et al.* (2007) Chen *et al.* (2007) Liu *et al.* (2005b)
Glomus geosporum Glomus etunicatum	*Pteris vittata* L.	Wu *et al.* (2009)
Glomus mosseae, Glomus intraradices, Glomus etunicatum	*Pityrogramma calomelanos* L. *Tegetes erecta* L. *Melastoma malabathricum* L.	Jankong and Visoottiviseth (2008)

(Continued)

Table 6.4. Continued.

Augmented microbes	Plants	Reference
Acaulospora morrowiae	*Zea mays* L.	Wang *et al.* (2007)
Gigaspora margarita	*Pteris vittata* L.	Trotta *et al.* (2006)
Glomus caledonium,	*Pteris vittata* L.	Chen *et al.* (2006)
Glomus intraradices		
Glomus sp.	*Holcus lanatus* L.	Gonzalez-Chavez *et al.* (2002)
Hymenoscyphus ericae	*Calluna vulgaris* L. Hull.	Sharples *et al.* (2000)
FUNGI		
Aspergillus niger,	Soil	Mukherjee *et al.* (2010)
Aspergillus nidulans		Maheswari and Murugesan (2009)
Trichoderma harzianum,	*Eucalyptus globulus* Labill.	Arriagada *et al.* (2009)
Trametes versicolor		
Ulocladium sp.,	Soil	Edvantoro *et al.* (2004)
Penicillium sp.		

phase, and this could potentially solubilize and mobilize As held within or sorbed on the surface of the iron oxides (Benner *et al.*, 2002; Cummings *et al.*, 1999; Rowland *et al.*, 2007). However, some authors (Kocar *et al.*, 2006; Tufano *et al.*, 2008b) have observed that the Fe(III) reduction is also likely to form secondary iron phases at root surfaces of plants which are having a potential to absorb As. Wang and Zhou (2009) indicated that both microbial and chemical reductions of iron plaque caused slow As release from iron plaque to aqueous phases. However, microbial iron reduction induced the formation of more crystalline iron minerals, leading to As sequestration. The concentration of As in iron plaque was 170 mg kg^{-1} as compared to the soil As (42 mg kg^{-1}) demonstrating that the iron plaque formed naturally on the rice root surface could accumulate As.

6.4.8 *Microbes-As interaction*

Tolerance and adaptation of microorganisms to As are common phenomena. Presence of tolerant fungi and bacteria in As rich environment have frequently been observed (Butt and Rehmann, 2011; Cavalca *et al.*, 2010; Srivastava *et al.*, 2011; Srivastava *et al.*, 2010; Su *et al.*, 2010b; Xu *et al.*, 2008). The increased abundance of tolerant microbes can be due to genetic changes, physiological adaptations involving no alterations in the genotype or replacement of As sensitive species with microbial species that already tolerant to As.

Metal resistance in bacteria is often encoded by genes located on plasmids. Bacteria develop As resistance mechanisms through the *Ars* operon genes (Kaur *et al.*, 2011; Muckopadhyay *et al.*, 2002; Oremland *et al.*, 2005; Silver and Phung, 2005). As stated earlier, typical *Ars* operon contains either (*Ars* RBC) or five (*Ars*RDABC) genes that generally transcribe as a single unit (Rosen, 1999). *Ars*R is a repressor that binds the promoter region and regulates the *Ars* operon. *Ars*B is a membrane-located transport protein that can pump As(III) out of cells using proton motive force. *Ars*C was shown to be a cytoplasmic As(V) reductase, whereas *Ars*A is an As(III) activated ATPase (Zhou *et al.*, 2000).

As(III) is generally considered more toxic that As(V) for most of the plants. However, plants vary in their sensitivity or resistance to As (Meharg and Hartley-Whitaker, 2002). Rhizobacteria encounter As(V) and As(III) in soil solutions before they enter the root and oxidizing As(III) from soil solutions might help the plant to grow on As-contaminated soils and thus lower the As(III) toxicity. The presence of numerous As-resistant bacteria with plant growth promoting characteristics e.g., siderophores, IAA, and ACC deaminase, could potentially support plant growth in As- contaminated areas and reduce stress symptoms/toxicity (Zaidi *et al.*, 2006).

6.5 STRATEGIES FOR BIOREMEDIATION OF ARSENIC

A successful bioremediation program usually requires the application of strategies customized for the specific environmental conditions of the contaminated sites. Further, bioremediation utilizes the metabolic potential of microorganisms to clean up contaminated environments. Bioremediation is carried out in non-sterile open environments that contain a variety of organisms that affect the process. Microbial and ecological information are useful for the development of strategies to improve bioremediation and for evaluating its consequences. Strategies involving the addition of seeded cultures (bioaugmentation) or the addition of nutrients (biostimulation), hold the promise of fostering bioremediation/detoxification degradation rates (Atlas, 1995; Jimenez *et al.*, 2006). The impending gap between laboratory trials and on-field studies are due to several factors influencing the remediation process, like strain selection, indigenous microbial ecology, environmental constraints and the procedures used for *in-situ* remediation. The fate and transport of As in soil and groundwater also depends on the chemical form and speciation of As. Examples of bacteria and fungi, which have been investigated for environmental remediation of As contamination, are shown in Tables 6.2 and 6.3.

6.5.1 *Screening and selection of suitable microbes*

Feasibility studies are the pre-requisite for any planned bioremediation intervention that includes screening followed by tailoring of a microbial formulation for a particular site. The initial screening/selection step is based on microbial metabolism, which is functionally active and persistent under the desired environmental conditions. The best approach for selecting competent microbes should be based on the prior knowledge of the microbial communities inhabiting the target site (Thomson *et al.*, 2005; Vander Gast *et al.*, 2004). From the applied prospective, use of consortia of pure cultures for the bioremediation is more advantageous which could have metabolic robustness in the field application (Ledin, 2000). Corsini *et al.* (2011) isolated As–resistant bacteria (*Bacillus cereus, Pseudomonas azotoformans* and *Phodococcus erythropolis*) from citrate-amended soil, which were able to reduce 2 mmol As L^{-1} in liquid culture. In a separate study, five bacterial isolates transformed arsenate to arsenite and volatile methyl arsines. These strains belonged to the *Proteus, Escherichia, Flavobaterium, Corynebacterium* and *Pseudomonas* genera (Ordonez *et al.*, 2005; Shariatpanahi *et al.*, 1981). As given in Section 3.2.4., strain *R. palustris* was capable of forming of a number of methylated intermediates of As(III), with trimethyl arsine as the end product (Qin *et al.*, 2006). Many fungi have been found capable of As accumulation and/or volatilization such as *Scopulariopsis brevicaulis* (Gosio, 1892), *Phalolus schweinitzii* (Pearce, 1998), *Fusarium oxysporum* (Granchinho *et al.*, 2002), *Sinorhizobium melitoti* (300 mg L^{-1} As) (Carrasco *et al.*, 2005). As uptake in *Aspergillus candidus* were measured as 11.17, 4.09, and 8.00 mg g^{-1} on day 3, 6 and 9, respectively, when exposed to initial 50 mg L^{-1} arsenate (Vala, 2010). The mean percent removal as flux of biovolatilized As ranged from 3.71–29.86% in *Trichoderma* sp., *Necosmospora* sp., *and Rhizopus* sp. can be effectively used for the bioremediation of As contaminated agricultural soils (Srivastava *et al.*, 2011). Su *et al.* (2010a) have found three fungal strains *Trichoderma asperellum* SM-12F1, *Penicillium janthinellum* SM-12F4, and *Fusarium oxysporum* CZ-8F1 that are highly capable of As accumulation and volatilization. *Trichoderma asperellum* SM-12F1, *Penicillium janthinellum* SM-12F4, and *Fusarium oxysporum* CZ-8F1 were exposed to 50 mg L^{-1} of As(V), and the biotransformation of As and the concomitant variance of Eh and pH of media was studied after cultivation for 2 or 3 days. The arsenate added to the media had been completely changed into arsenite, whilst arsenate was predominate in fungal cells with concomitantly little arsenite during cultivation. After 15 days, the total As (t-As) content was the highest (as 41.5 µg mg^{-1}) in cells of

P. janthinellum SM-12F4 according to the quantitative analysis of As speciation in cultures. This shows the widespread ability of diverse microbes to act on As and further its detoxification.

6.5.2 *Identification and manipulation of a functionally active microbial population*

The study of metal resistance in microbes has led to the discovery of many metal specific genetic models. In addition, molecular approaches are more frequently being used to study microbial communities in metal contaminated environments. Advantages of using molecular approaches include enhanced assessment of specific microbial populations in contaminated environments (Rochelle *et al.*, 1991) and increased sensitivity in monitoring specific microbial populations during remediation efforts (Mateos *et al.*, 2006). There are different mechanisms involved amongst the microbes for As tolerance and removal. These conversion mechanisms are mediated by the genetic determinants mostly associated with plasmids (Mobley and Rosen, 1982). Having established the plasmid mediated nature of As resistance and detoxification, an analysis of the genetic diversity of the *ars* determinants would provide a better understanding of the adaptive responses of the particular microbe to As. Several approaches to investigate the distribution of divergence of *ars* determinants in natural environments have been used. Baker and Brooks (1989) used a probe to detect homologous determinants by DNA-DNA hybridization, followed by more extensive hybridization studies using a series of *ars* probes against DNA isolated from bacteria.

6.5.3 *Specific functions of microbes*

Microbial transformation of As has great implication on the chemical cycling in the environment, because different forms of As vary in solubility, mobility, bioavailability and toxicity (Smedley and Kinniburgh, 2002). Microbes affect As reduction and oxidation, methylation and demethylation and sorption and desorption in soils. As a result, microbes have developed different detoxification strategies to withstand the growth restriction under As stress. Microbial As(V) reducing mechanisms can be classified into two types: detoxification and dissimilation. A variety of bacteria are reported to reduce As(V) to As(III). Occasionally, bacteria take a large quantity of As(V) during the process of phosphate uptake and in order to remove As(V) toxicity, they convert As(V) to As(III), which has a high mobility and is expelled out of the cells (Rosen, 2002). Such microbial detoxification of As occurs under both aerobic and anaerobic conditions, but is not coupled to the energy generation. Although, some anaerobes can gain energy to support growth and cell function by coupling As(V) reduction to oxidation of soil organic matter. Oremland and Stolz (2003) had reported 16 As(V)-respiring prokaryotes. Later, Stolz *et al.* (2010) isolated more than 50 phylogenetically diverse As(III)-oxidizing strains distributed among more than 13 genera from various environments. These include chemolithoautotrophic As(III) oxidizers, which use the energy and reducing power from As(III) oxidation during CO_2 fixation and cell growth under both aerobic (Santini *et al.*, 2000) and nitrate reducing conditions (Oremland *et al.*, 2002; Rhine *et al.*, 2006).

Bacteria have also shown to increase the As mobility in nature through reductive dissolution of Fe(III)-oxides (Grantham *et al.*, 1997; Lee *et al.*, 2009; Lovley, 1993). Lovley and Coates (1997) suggested that Fe-reducing bacteria dissolve Fe(III) to aqueous Fe(II) and subsequently As associated with Fe minerals was leached out in anaerobic environments.

6.5.4 *Genetically engineered (GE) bacteria*

Application of GE bacteria is now widely suggested for remediation of various sites contaminated with metal pollutants. A combination of microbiological and ecological knowledge, biochemical mechanisms and field engineering designs are essential elements for successful bioremediation of metal contaminated sites using engineered bacteria. The GE bacteria have higher detoxification capacity and have been demonstrated successfully for the degradation of various pollutants under defined conditions (Barac *et al.*, 2004). Various bacterial strains of *Ralstonia eutropha* (its mtb gene for remediation of Cd^{2+}), *Mycobacterium marimum* (its MerH gene for remediation of

Hg^{2+}), *E. coli, Sphingomonas desiccabilis, Bacillus idriensis* (its ArsM gene for volatilization of As) etc. have genes that would help in bioremediation of specific metals (Valls *et al.*, 2000; Schue *et al.*, 2009; Liu *et al.*, 2011).

Bioremediation of As from contaminated sites by using GE bacteria could be an option to transform arsenic to its non-toxic forms (Valls and De Lorenzo, 2002). Expression of phytochelatin (PC) synthase of *Arabidopsis thaliana* (ATPCS) in *E. coli* cells (Sauge-Merle *et al.*, 2003) increased the accumulation of intracellular arsenic by 50-fold by sequestering arsenite in a non-toxic form and competing with the arsenic efflux transporter. In another scenario, *E. coli* expressing arsenite S-adenosylmethionine methyltransferase gene *(arsM)* cloned from (*Rhodopseudomonas pulustris*) has been found to methylate toxic i-As to less-toxic volatile trimethylarsine (Cullen and Bentley, 2005; Qin *et al.*, 2006; 2009). Recently Liu *et al.* (2011) demonstrated that As could be removed through volatilization by GE bacteria, showing *arsM* genes. They overexpressed *arsM* genes in *Sphingomonas desiccabilis* and *Bacillus idreinsis*, which resulted in 10-fold increase in methylated As release compared to their wild types. The transcriptional regulator *arsR* present in the bacterial *ars* operon can be considered an As-specific metallothionein like protein endowed with an unambiguous binding site for arsenite (Paez-Espino *et al.*, 2009). These results open the possibility of designing As-specific bioadsorbants by merging dedicated transport systems with naturally occurring or evolved intracellular As-binding polypeptides.

6.5.5   *Enhancement of bioremediation by use of surfactants*

Bio-surfactants are microbial compounds that exhibit high surface and emulsifying activity. The major classes of bio-surfactants include glycolipids, lipopeptides and lipoproteins, phospholipids, fatty acids, polymeric surfactants, and particulate surfactants. Bio-surfactants production, structure and their various natural and industrial uses and their role in pollutant removal have been extensively reviewed (Banat, 1995; Banat *et al.*, 2000; Cameotra and Bollag, 2003; Eliora and Rosenberg, 2001; Singh and Cameotra, 2004). The efficiency of bio-surfactants for stimulating As detoxification/removal is uncertain due to specificity observed between biosurfactant and microorganisms. A strategy suitable for effective As-remediation would be to stimulate biosurfactants produced by indigenous microbial population or use of commercial bio-surfactants produced by biological organisms found to be already present at the contaminated site (Singh and Cameotra, 2004).

6.5.6   *Priming and encapsulation*

The major concern in bioremediation is the survival along with the soil colonization of the microbial inoculants consisting of pure cultures or consortia. Bioremediation very often fails because of the low survival rate of the inoculated microorganisms. The capacity of microorganisms to proliferate in the soil after bioaugmentation is important as their performance, activities and persistence are related to the microorganisms characteristics and inoculation procedure (Thomson *et al.*, 2005). For the success of bioaugmentation, the previous conditioning of the microorganisms before inoculation needs improvement. One method is 'priming' or 'activated soil' strategy. This consists of pre-inoculation of the microbial inoculums in sterilized soil and to be incubated for 7–15 days before bioaugmentation (Gentry *et al.*, 2004). The advantages of this method are: (i) pool of useful complementary microorganisms, (ii) pool of cultivable and non-cultivable microorganisms, (iii) no steps of extraction and culture of microbes and (iv) a better microbial survival rate since soil serves as a carrier for microbes.

Another method is immobilizing microorganisms into carriers or 'encapsulation'. For example, alginate, clay, peat etc., protect microbes against biotic (Cassidy *et al.*, 1996; Da Silva and Alvarez, 2010; Gentry *et al.*, 2004; McLoughlin, 1994; Tyagi *et al.*, 2011) and abiotic environmental stress such as toxicity of metals (Braud *et al.*, 2007). Alginate encapsulated plant growth promoting bacteria (PGPB) can be lyophilized and stirred at high density for the extended periods of time, making this method appealing for field-scale and commercial use (Reed and Glick, 2005). Alginate beads are also thought to offer protection to the PGPB in harsh environment by acting as a time-release coating that slowly disintegrates and releases PGPB to the germinating plants (Bashan *et al.*, 2002). While studying the capacity of alignate encapsulation of PGPB for revegetation in mine tailing, Grandlic *et al.* (2009) observed that encapsulation was an effective way to inoculate PGPB. However, the feasibility of this approach at the field scale is limited by the amount of carriers required to reach acceptable degradation (Owsianiak *et al.*, 2010). The field trials are required to evaluate and validate bioaugmentation with immobilized microorganism (encapsulated) as a remediation strategy for As-contaminated soils and to determine the required amount of carriers to be introduced into soil (Leung *et al.*, 1995). Examples of bioaugmentation by microbes in soil for some plants are shown in Table 6.4 for the remediation of As.

6.6   CONCLUSIONS

In-situ bioremediation of As contaminated soils shows great potential for future developments due to its environmental compatibility. It relies on microbial activities to reduce, mobilize or immobilize As through biosorption, biomethylation, complexation, oxidation-reduction, and bio-volatilization processes. Bioremediation is widely accepted and cost-effective applied technology. Full-scale field demonstrations are required along with assessment of cost-effective analysis. Strategies like use of bio-surfactants, and other suitable nutrients and organics to reduce As bioavailability should be investigated. The As tolerance, accumulation and biotransformation capabilities of microbes may be improved through genetic engineering strategies. One major limitation of *in-situ* bioremediation is its application in natural environments, where uncharacterized microorganisms exist. Further, the environmental conditions are also varying with respect to habitats, climate and hydro-geo dynamics. Compilation of database, taking the results of As-contaminated sites and its bioremediation techniques, including the ecological risk assessment is also essential for developing a site-specific bioremediation scheme.

ACKNOWLEDGEMENTS

Authors are thankful to Director, CSIR-National Botanical Research Institute, for providing institutional support. Financial supports from Department of Biotechnology, Government of India (BT/PR13147/BCE/08/780/2009) and Council of Scientific and Industrial Research are thankfully acknowledged. Shubhi Srivastava is thankful to CSIR for providing Senior Research fellowship (CSIR-SRF).

REFERENCES

Abdulsalam, S., Bugaje, I.M., Adefila, S.S. & Ibrahim S.: Comparison of biostimulation and bioaugmentation for remediation of soil contaminated with spent motor oil. *Int. J. Environ. Sci. Tech.* 8 (2011), pp. 187–194.

Abedin, M.J., Feldmann, J. & Meharg, A.A.: Uptake kinetics of arsenic species in rice plants. *Plant Physiol.* 128 (2002), pp. 1120–1128.

Adeyemi, A.O.: Bioaccumulation of arsenic by Fungi. *Amer. J. of Environ. Sci.* 5:3 (2009), pp. 364–370.

Adriano, D.: *Trace elements in terrestrial environments: Biogeochemistry, bioavailability, and risk of metals.* Springer-Verlag, New York, 2001.

Adriano, D.C., Page, A.L., Elseewi, A.A., Chang, A.C. & Straughan, I.: Utilization and disposal of fly ash and other residues in terrestrial ecosystems: A review. *J. Environ. Qual.* 9 (1980), pp. 333–344.

Agely, A.A., Sylvia, D.M. & Ma, L.Q.: Mycorrhizae increase arsenic uptake by the hyperaccumulator chinese brake fern (*Pteris vittata* L.). *J. Environ. Qual.* 34 (2005), pp. 2181–2186.

Aggangan, N.S., Dell, B. & Malajczuk, N.: Effects of chromium and nickel on growth of the ectomycorrhizal fungus *Pisolithus* and formation of ectomycorrhizas on *Eucalyptus urophylla* S.T. Blake. *Geoderma* 84:1–3 (1998), pp. 15–27.

Ahmann, D., Roberts, A.L., Krumholz, L.R. & Morel, F.M.: Microbes grow by reducing arsenic. *Nature* 371 (1994), p. 750.

Ahumada, I., Escudero, P., Castillo, G., Carrasco, A., Ascar, L. & Fuentes, E.: Use of sequential extraction to assess the influence of sewage sludge amendment on metal mobility in Chilean soils. *J. Environ. Monitor.* 6 (2004), pp. 327–334.

Aksornchu, P., Prasertsan, P. & Sobhon, V.: Isolation of arsenic-tolerant bacteria from arsenic-contaminated soil. *Songklanakarin J. Sci. Technol.* 30 (Suppl. 1) (2008), pp. 95–102.

Alam, M.G.M., Snow, E.T. & Tanaka, A.: Arsenic and heavy metal contamination of vegetables grown in Samta village. *Bangladesh Sci. Tot. Environ.* 308: (2003), pp. 83–96.

Alexander, M.: *Introduction to soil microbiology.* 2nd ed., John Wiley and Sons, Inc., New York, NY, 1977.

Anderson, G.L., Love, M. & Zeider, B.K.: Metabolic energy from arsenite oxidation in *Alcaligenes faecalis. Journal De Physique. IV: JP* 107:I (2003), pp. 49–52.

Andreae, M.O.: Organo arsenic compounds in the environment. In: P.J. Craig (ed): *Organometallic compounds in the environment.* Longman, Harlow, 1986, pp. 198–228.

Arriagada, C., Aranda, E., Sampedro, I., Garcia-Romera, I. & Ocampo, J. A.: Contribution of the saprobic fungi *Trametes versicolor* and *Trichoderma harzianum* and the arbuscular mycorrhizal fungi *Glomus deserticola* and *G. claroideum* to arsenic tolerance of *Eucalyptus globulus. Biores. Technol.* 100:2 (2009), pp. 6250–6257.

Atlas, R.M.: Bioremediation. *Chem. Eng. News* 73 (1995) pp. 32–42.

Aurelius, L.: Investigation of arsenic contamination of groundwater occurring near Knott, Texas. Austin, Texas. Texas Department of Agriculture, TX, 1988.

Bai, R.S. & Abraham, T.E.: Studies on chromium (VI) adsorption–desorption using immobilized fungal biomass. *Biores. Technol.* 87: (2003), pp. 17–26.

Bais, H.P., Weir, T.L., Perry, L.G., Gilroy, S. & Vivanco, J.M.: The role of root exudates in rhizosphere interactions with plants and other organisms. *Ann. Rev. Plant. Biol.* 57 (2006), pp. 233–266.

Baker, A.J.M. & Brooks, R.R.: Terrestrial higher plants which hyperaccumulate metallic elements- a review of their distribution, ecology, and phytochemistry. *Biorecovery* I (1989), pp. 81–126.

Banat, I.M.: Characterization of biosurfactants and their use in pollution removal-state of art (review). *Acta Biotechnol.* 15: (1995), pp. 251–267.

Banat, I.M., Makkar, R.S. & Cameotra, S.S.: Potential commercial application of microbial surfactants. *Appl. Microbiol. Biotechnol.* 53 (2000), pp. 495–508.

Barac, T., Taghavi, S., Borremans, B., Provoost, A., Oeyen, L., Colpaert, J.V., Vangronsveld, J. & Van der Lelie, D.: Engineered endophytic bacteria improve phytoremediation of water-soluble, volatile, organic pollutants. *Nat. Biotech.* 22 (2004), pp. 583–588.

Barua, A., Gupta, S.D., Mridha, M.A.U. & Bhuiyan, M.K.: Effect of arbuscular mycorrhizal fungi on growth of *Gmelina arborea* in arsenic-contaminated soil. *J. Forest. Res.* 21:4 (2010), pp. 423–432.

Bashan, Y., Hernandez, J.P., Leyva, L.A. & Bacilio, M.: Alginate microbeads as inoculant carriers for plant growth-promoting bacteria. *Biol. and Ferti. Soils* 35 (2002), pp. 359–368.

Batten, K.M., Scow, K.M., Davies, K.F. & Harrison, S.P.: Two invasive plants alter soil microbial community composition in serpentine grasslands. *Biol. Inv.* 8 (2006), pp. 217–230.

Benner, S.G., Hansel, C.H. Wielinga, B.W. Barber, T.M. & Fendorf, S.: Reductive dissolution and biomineralization of iron hydroxide under dynamic flow conditions. *Environ. Sci. Technol.* 36 (2002), pp. 1705–1711.

Bentley, R. & Chasteen, T.G.: Microbial methylation of metalloids: Arsenic, antimony, and bismuth. *Microbiol. Mol. Biol. Rev.* 66:2 (2002), pp. 250–271.

Berg, M., Tran, H.C., Nguyen, T.C., Pham, H.V., Shertenleib, I. & Giger, W.: Arsenic contamination of groundwater and drinking water in Vietnam: A human health threat. *Environ. Sci. Technol.* 35:13 (2001), pp. 2621–2626.

Bhattacharya, P., Chatterjee, D. & Jacks, G.: Occurrence of arsenic contaminated groundwater in alluvial aquifers from Delta Plains, Eastern India: Options for safe drinking water supply. *Internat. J. Wat. Resour. Manag.* 13 (1997), pp. 79–92.

Bobrowicz, P., Wysocki, R., Owsianik, G., Goffeau, A. & Ulaszewski, S.: Isolation of three contiguous genes, ACR1, ACR2 and ACR3, involved in resistance to arsenic compounds in the yeast *Saccharomyces cerevisiae. Yeast* 13 (1997), pp. 819–828.

Borch, T., Kretzschmar, R., Kappler, A., Van Cappelen, P., Ginder-Vogel, M., Voegelin, A. & Campbell, K.: Biogeochemical redox processes and their impact on contaminant dynamics. *Environ. Sci. Technol.* 44: (2010), pp. 15–23.

Brammer, H.: Mitigation of arsenic contamination in irrigated paddy soils in south and south East Asia. *Environ. Int.* 35 (2009), pp. 856–863.

Braud, A., Jézéquel, K. & Lebeau, T.: Impact of substrates and cell immobilization on siderophore activity by Pseudomonads in a Fe and/or Cr, Hg, Pb containing-medium. *J. Hazard. Mater.* 144 (2007), pp. 229–239.

Brierley, J.A. & Brierley C.L.: Present and future commercial applications of biohydrometallurgy. *Hydrometallurgy* 59 (2001), pp. 233–239.

Bruckner, M.: Comparison of *in-situ* bioremediation technologies. *EM: Air and Waste Management Association's Magazine for Environmental Managers* (June 2011), pp. 20–23.

Bruhn, D.F., Li, J., Silver, S., Roberto, F. & Rosen, B.P.: The arsenical resistance operon of IncN plasmid R46. *FEMS Microbiol. Lett.* 139 (1996), pp. 149–153.

Buckova, M., Godocikova, J. & Polek, B.: Responses in the mycelial growth of *Aspergillus niger* isolates to arsenic contaminated environments and their resistance to exogenic metal stress. *J. Basic Microbiol.* 47 (2007), pp. 295–300.

Butt, A.S. & Rehman, A.: Isolation of arsenite-oxidizing bacteria from industrial effluents and their potential use in wastewater treatment. *World J. Microbiol. Biotechnol.* 27:10 (2011), pp. 2435–2441.

Cai, L., Liu, G., Rensing, C. & Wang, G.: Genes involved in arsenic transformation and resistance associated with different levels of arsenic-contaminated soils. *BMC Microbiol.* 9:4 (2009).

Cameotra, S.S. & Bollag, J.M.: Biosurfactant enhanced bioremediation of PAH. *CRC Crit. Rev. Environ. Sci. Technol.* 30 (2003), pp. 111–126.

Campbell, K.M., Root, R., O'Day, P.A. & Hering, J.G.: A gel probe equilibrium sampler for measuring arsenic porewater profiles and sorption gradients in sediments: II. Field application to Haiwee Reservoir sediment. *Environ. Sci. Technol.* 42 (2008), pp. 504–510.

Cánovas, D., Cases, I. & de Lorenzo, V.: Heavy metal tolerance and metal homeostasis in *Pseudomonas putida* as revealed by complete genome analysis. *Environ. Microbiol.* 5 (2003), pp. 1242–1256.

Canovas, D., Vooijs, R., Schat, H. & de Lorenzo, V.: The role of thiol species in the hypertolerance of *Aspergillus* sp. P37 to arsenic. *J. Biol. Chem.* 279:49 (2004), pp. 51,234–51,240.

Carrasco, J.A., Armario, P., Pajuelo, E., Burgos, A., Caviedes, M.A., López, R., Chamber, M.A. & Palomares, A.J.: Isolation and characterization of symbiotically effective *Rhizobium* resistant to arsenic and heavy metals after the toxic spill at the Aznalco'llar pyrite mine. *Soil Biol. Biochem.* 37 (2005), pp. 1131–1140.

Cassidy, M.B., Lee, H. & Trevors, J.T.: Environmental applications of immobilized microbial cells: a review. *J. Ind. Microbiol.* 16 (1996), pp. 79–101.

Caudill, M.: *The effects of arsenic on Thiobacillus ferrooxidans.* Masters Thesis, Columbia University Earth and Environmental Engineering Department, New York, NY, 2003.

Cavalca, L., Zanchi, R., Corsini, A., Colombo, M., Romagnoli, C., Canzi, E. & Andreoni, V.: Arsenic resistant bacteria associated with roots of the wild *Cirsium arvense* (L.) plant from an arsenic polluted soil, and screening of potential plant growth promoting characteristics. *System. Appl. Microbiol.* 33:3 (2010), pp. 154–164.

Ceribasi, H.I. & Yetis, U.: Biosorption of Ni(II) and Pb(II) by *Phanaerochate chrysosporium* from a binary metal system-kinetics. *Water SA.* 27:1 (2001), pp. 15–20.

Cernansky, S., Urik, M., Sevc, J. & Khun, M.: Biosorption and biovolatilization of arsenic by heat-resistant fungi. *Env. Sci. Pollut. Res.* 14(1) (2007), pp. 31–35.

Cernansky, S., Kolencik, M., Sevc, J., Urik, M. & Hiller, E.: Fungal volatilization of trivalent and pentavalent arsenic under laboratory conditions. *Biores. Techhnol.* 100 (2009), pp. 1037–1040.

Chakraborti, D., Rahman, M.M., Paul, K., Chowdhury, U.K., Sengupta, M.K., Lodh, D., Chanda, C.R., Saha, K.C. & Mukherjee, S.C.: Arsenic calamity in the Indian subcontinent: what lessons have been learned? *Talanta* 58 (2002), pp. 3–22.

Chakraborti, D., Sengupta, M.K., Rahman, M.M., Ahamed, S., Chowdhury, U.K., Hossain, M.A., Mukherjee, S.C., Pati, S., Saha, K.C., Dutta, R.N. & Zaman, Q.Q.: Groundwater arsenic contamination and its health effects in the Ganga-Meghna-Brahmaputra plain. *J. Environ. Monitor.* 6 (2004), pp. 74–83.

Challenger, F.: Biological methylation. *Chem. Rev.* 36 (1945), pp. 315–361.

Challenger, F.: Biosynthesis of organometallic and organometalloidal compounds. In: F.E. Brinckman & J.M. Bellama (eds): *Organometals and organometalloids. Occurrence and fate in the Environment.* American Chemical Society, Washington D.C. (1978), pp. 1–22.

Chang, J.-S., Yoon, I.-H. & Kim, K.W.: Isolation and ars detoxification of arsenite-oxidizing bacteria from abandoned arsenic-contaminated mines. *J. Microbiol. Biotechnol.* 17:5 (2007), pp. 812–821.

Chang, J-S., Kim, Y.-H., & Kim, K.-W.: The ars genotype characterization of arsenic-resistant bacteria from arsenic-contaminated gold–silver mines in the Republic of Korea. *Appl. Microbiol. Biotechnol,* 80 (2008), pp. 155–165.

Chatain, V., Bayard, R., Sanchez, F., Moszkowicz, P. & Gourdon, R.: Effect of indigenous bacterial activity on arsenic mobilization under anaerobic conditions. *Environ. Int.* 31:2 (2005), pp. 221–226.

Chatterjee, A., Das, D., Mandal, B.K., Roy Chowdhury, T., Samanta, G. & Chakraborti, D.: Arsenic in ground water in six districts of West Bengal, India: The biggest arsenic calamity in the world, Part I. Arsenic species in drinking water and urine of the affected people. *Analyst* 120 (1995), pp. 645–650.

Chen, B.D., Zhu, Y.-G. & Smith, F.A.: Effects of arbuscular mycorrhizal inoculation on uranium and arsenic accumulation by Chinese brake fern (*Pteris vittata* L.) from a uranium mining-impacted soil. *Chemosphere* 62 (2006), pp. 1464–1473.

Chen, B.D., Zhu, Y.-G., Duan, J., Xiao, X.Y. & Smith, S.E.: Effects of the arbuscular mycorrhizal fungus *Glomus mosseae* on growth and metal uptake by four plant species in copper mine tailings. *Environ. Pollut.* 147 (2007), pp. 374–380.

Chitpirom, K., Akaracharanya, A., Tanasupawat S., Leepipatpiboon N. & Kim, K.W.: Isolation and characterization of arsenic resistant bacteria from tannery wastes and agricultural soils in Thailand. *Annals Microbiol.* 59:4 (2009), pp. 649–656.

Chopra, B.K., Bhat, S., Mikheenko, I.P., Xu, Z., Yang, Y., Luo, X., Chen, H., Zwieten, L.V., Lilley R.McC. & Zhang, R.: The characteristics of rhizosphere microbes associated with plants in arsenic-contaminated soils from cattle dip sites. *Sci. Tot. Environ.* 378 (2007), pp. 331–342.

Clement, R., Walker, D.J. & Roig, A.: Heavy metal bioavailability in a soil affected by mineral sulphides contamination following the mine spillage at Aznalcollar (Spain). *Biodegra.* 14 (2003), pp. 199–205.

Cooper, P.A. & Ung, Y.T.: Effect of water repellents on leaching of CCA from treated fence and deck units-An update. IRG/WP 97-50086, International Research Group, Stockholm, Sweden, 1997.

Corsini, A., Cavalca, L., Zaccheo, P., Crippa, L. & Andreoni, V.: Influence of microorganisms on arsenic mobilization and speciation in a submerged contaminated soil: Effects of citrate. *Appl. Soil Ecol.* 49:1 (2011), pp. 99–106.

Cosgrove, L., Mac Geechan, P.L., Handley, P.S. & Robson, G.D.: Effect of biostimulation and bio-augmentation on degradation of polyurethane buried in soil. *Appl. Environ. Microbiol.* 76:3 (2010), pp. 810–819.

Couto, M.N.P.F.S., Monteiro, E. & Vasconcelos, M.T.S.D.: Mesocosm trials of bioremediation of contaminated soil of a petroleum refinery: Comparison of natural attenuation, biostimulation and bioaugmentation. *Environ. Sci. Pollut. Res.* 17:7 (2010), pp. 1339–1346.

Cox, D.P. & Alexander, M.: Effect of phosphate and other anions on trimethylarsine formation by *Candida humicola*. *Appl. Microbiol.* 25 (1973), pp. 408–413.

Cullen, W.R. & Bentley, R.: The toxicity of trimethylarsine: an urban myth. *J. Environ. Moni.* 7 (2005), pp. 11–15.

Cullen, W.R. & Reimer K.J.: Arsenic speciation in the environment. *Chem. Rev.* 89: (1989), pp. 713–764.

Cummings, D.E., Caccavo Jr, F., Fendorf, S. & Rosenzweig, R.F.: Arsenic mobilization by the dissimilatory Fe(III)-reducing bacterium *Schewanella alga* BrY. *Environ. Sci. Technol.* 33 (1999), pp. 723–729.

Da Silva, M.L.B. & Alvarez, P.J.J.: Bioaugmentation. In: K.N. Timmis (ed): *Handbook of hydrocarbon and lipid microbiology*; Springer-Verlag, Berlin-Heidelberg, Germany, 2010, pp. 4531–4544.

Das, H.K., Mitra, A.K., Sengupta, P.K., Hossain, A., Islam, F. & Rabbani, G.H.: Arsenic concentrations in rice, vegetables, and fish in Bangladesh: a preliminary study. *Environ. Int.* 30 (2004), pp. 383–387.

Davenport, J.R. & Peryea, F.J.: Phosphate fertilizers influence leaching of lead and arsenic in a soil contaminated with lead and arsenic in a soil contaminated with lead arsenate. *Water Air Soil Pollut.* 57:58 (1991), pp. 101–110.

Del Razo, L.M., Arellano, M.A. & Cebrián, M.E.: The oxidation states of arsenic in well water from a chronic arsenicism area of northern Mexico. *Environ. Pollut.* 64 (1990), pp. 143–153.

Demirbas, A.: Heavy metal bioaccumulation by mushrooms from artificially fortified soils. *Food Chem.* 74 (2001), pp. 293–301.

D'Imperio, S., Lehr, C.R., Breary, M. & Mac Dermott, T.R.: Autecology of an arsenite chemolithotroph: sulfide constraints on function and distribution in a geothermal spring. *Appl. Environ. Microbiol.* 73 (2007), pp. 7067–7074.

Dixit, S. & Hering, J.G.: Comparison of arsenic(V) and arsenic(III) sorption onto iron oxide minerals: implications for arsenic mobility. *Environ. Sci. Tech.* 37 (2003), pp. 18:4182.

Dong, Y., Zhu, Y.-G., Smith, F.A., Wang, Y. & Chen, B.: Arbuscular mycorrhiza enhanced arsenic resistance of both white clover (*Trifolium repens* Linn.) and ryegrass (*Lolium perenne* L.) plants in an arsenic-contaminated soil. *Environ. Pollut.* 155 (2008), pp. 174–181.

Edvantoro, B.B., Naidu, R., Megharaj, M., Merrington, G. & Singleton, I.: Microbial formation of volatile arsenic in cattle dip site soils contaminated with arsenic and DDT. *Appl. Soil Ecol.* 25 (2004), pp. 207–217.

Eguchi, N., Kuroda, K. & Endo, G.: Metabolites of arsenic induced tetraploids and mitotic arrest in cultured cells. *Arch. Environ. Contam. Toxicol.* 32 (1997), pp. 141–145.

Eliora, Z.R. & Rosenberg, E.: Natural role of biosurfactants. *Environ. Microbiol.* 3 (2001), pp. 229–236.

Fendorf, S., La Force, M.J. & Li, G.: Temporal changes in soil partitioning and bioaccessibility of arsenic, chromium, and lead. *J. Environ. Qual.* 33 (2004), pp. 2049–2055.

Filippi, M., Goliáš, V. & Pertold, Z.: Arsenic in contaminated soils and anthropogenic deposits at the Mokrsko, Roudný, and Kašperské Hory gold deposits, Bohemian Massif (CZ). *Environ. Geol.* 45 (2004), pp. 716–730.

Gadd, G.M.: Bioremedial potential of microbial mechanisms of metal mobilization and immobilization. *Curr. Opin. Biotechnol.* 11 (2000), pp. 271–279.

Gadd, G.M.: Microbial influence on metal mobility and application for bioremediation. *Geoderma* 122 (2004), pp. 109–119.

Gao, S. & Barau, R.G.: Environmental factors affecting rates of arsine evolution from and mineralization of arsenicals in soil. *J. Environ. Qual.* 26 (1997), pp. 753–763.

Gentry, T.J., Rensing, C. & Pepper, I.L.: New approaches for bioaugmentation as a remediation technology. *Crit. Rev. Environ. Sci. Technol.* 34 (2004), pp. 447–494.

Ghosh, A.K., Bhattacharyya, P. & Pal, R.: Effect of arsenic contamination on microbial biomass and its activities. *Environ. Int.* 30 (2004), pp. 491–499.

Ghosh, M., Shen, J. & Rosen, B.P.: Pathways of As(III) detoxification in *Saccharomyces cerevisiae*. *PNAS* 96 (1999), pp. 5001–5006.

Ghosh, P., Rathinasabapathi, B. & Ma, L.Q.: Arsenic resistant bacteria solubilized arsenic in the growth media and increased growth hyperaccumulator *Pteris vittata* L. *Bioresour. Technol.* 102 (2011), pp. 8756–8761.

Gihring, T.M. & Banfield, J.F.: Arsenite oxidation and arsenate respiration by a new *Thermus* isolate, *FEMS Microbiol. Lett.* 204 (2001), pp. 335–340.

Gihring, T.M., Druschel, G.K., McKlesky, R.B., Hamers, R.J. & Banfield, J.F.: Rapid arsenite oxidation by *Thermus aquaticus* and *Thermus thermophilus*: field and laboratory investigations. *Environ. Sci. Technol.* (2001), pp. 35:3857.

Goldberg, S.: Competitive adsorption of arsenate and arsenite on oxide and clay minerals. *Soil Sci. Soc. Amer. J.* 66 (2002), pp. 413–421.

Gonzalez-Chavez, C., Harris, P.J., Dodd, J. & Meharg, A.A.: Arbuscular mycorrhizal fungi confer enhanced arsenate resistance on *Holcus lanatus*. *New Phytol.* 155 (2002), pp. 163–171.

Gosio, B.: Azione di alcune muffe sui composti fissi d'arsenico. *Bivista d'Igiene e Sanita Pubblica* III 8/9 (1892), pp. 201–230 and pp. 261–273.

Granchinho, S.C.R., Franz, C.M., Polishchuk, E., Cullen, W.R. & Reimer, K.J.: Transformation of arsenic(V) by the fungus *Fusarium oxysporum melonis* isolated from the alga *Fucus gardneri*. *Appl. Organomet. Chem.* 16 (2002), pp. 721–726.

Grandlic, C.J., Palmer, M.W., Maier, R.M.: Optimization of plant growth-promoting bacteria-assisted phytostabilization of mine tailings. *Soil Biol. Biochem.* 41 (2009), pp. 1734–1740.

Grantham, M.C., Dove, P.M. & DiChristina, T.J.: Microbially catalyzed dissolution of iron and aluminum oxyhydroxide mineral surface coatings. *Geochim. Cosmochim. Acta* 61 (1997), pp. 4467–4477.

Gunatilaka, A.A.L.: Natural products from plant-associated microorganisms: Distribution, structural diversity, bioactivity, and implications of their occurrence. *J. Natur. Product.* 69 (2006), pp. 509–526.

Hinsinger, P., Gobran, G.R., Gregory, P.J. & Wenzel, W.W.: Rhizosphere geometry and heterogeneity arising from root-mediated physical and chemical processes. *New Phytol.* 168 (2005), pp. 293–303.

Hofman, K., Hammer, E., Kohler, M. & Bruser, V.: Oxidation of triphenylarsine to triphenylarsineoxide by *Trichoderma harzianum* and other fungi. *Chemosphere* 44:4 (2001), pp. 697–700.

Hopenhayn-Rich, C., Biggs, M.L. & Smith, A.H.: Lung and kidney cancer mortality associated with arsenic in drinking water in Córdoba, Argentina. *Int. J. Epidemiol.* 27 (1998), pp. 561–569.

Huang, A., Teplitski, M., Rathinasabapathi, B. & Ma, L.Q.: Characterization of arsenic-resistant bacteria from the rhizosphere of arsenic hyperaccumulator *Pteris vittata* L. *Can. J. Microbiol.* 56 (2010), pp. 236–246.

Huang, Y.K., Lin, K.H., Chen, H.W., Chang, C.C., Liu, C.W., Yang, M.H. & Hsueh, Y.M.: Arsenic species contents at aquaculture farm and in farmed mouthbreeder (*Oreochromis mossambicus*) in blackfoot disease hyperendemic areas. *Food Chem. Toxicol.* 41 (2003), pp. 1491–1500.

Hutchinson, T.C., Aufreiter, S. & Hancock, R.G.V.: Arsenic pollution in the Yellowknife area from gold smelter activities. *J. Radioanal. Chem.* 71:1–2 (1982), pp. 59–73.

Ilyaletdinov, A.N. & Abdrashitova, S.A.: Autotrophic oxidation of arsenic by a culture of *Pseudomonas arsenitoxidans*. *Microbiol.* 50 (1981), pp. 197–204.

Jankong, P. & Visoottiviseth, P.: Effects of arbuscular mycorrhizal inoculation on plants growing on arsenic contaminated soil. *Chemosphere* 72 (2008) pp. 1092–1097.

Jenkins, R.O., Ritchie, A.W., Edmonds, J.S., Gössler, W., Molenat, N., Kühnelt, D., Harrington, C.F. & Sutton, P.G.: Bacterial degradation of arsenobetaine via dimethylarsinoylacetate. *Archiv. Microbiol.* 180 (2003), pp. 142–150.

Jimenez, B. & Wang, L.: Sludge treatment and management. In: Z. Ujang & M. Henze (eds): *Developing countries: Principles and engineering.* IWA Publishing, London, 2006, pp. 237–292.

Kadian, N., Gupta, A., Satya, S., Mehta, R.K. & Malik, A.: Biodegradation of herbicide (atrazine) in contaminated soil using various bioprocessed materials. *Bioresourc. Technol.* 99:11 (2008), pp. 4642–4647.

Kamilova, F., Kravchenko, L.V., Shaposhnikov, A.I., Azarova, T., Makarova, N. & Lugtenberg, B.: Organic acids, sugars and L-tryptophan in exudates of vegetables growing on stonewool and their effects on activities of rhizosphere bacteria. Mol. *Plant Microbe Intract.* 19 (2006), pp. 250–256.

Kandeler, E., Kampichler, C. & Horak, O.: Influence of heavy metals on the functional diversity of soil microbial communities. *Biol. and Fertil. Soils* 23 (1996), pp. 299–306.

Kaur, S., Kamli, M.R. & Ali, A.: Role of arsenic and its resistance in nature. *Can J. Microbiol.* 57 (2011), pp. 769–774.

Kavamura, V.N. & Esposito, E.: Biotechnological strategies applied to the decontamination of soils polluted with heavy metals. *Biotechnol. Adv.* 28 (2010), pp. 61–69.

Kocar, B.D., Herbel, M.J., Tufano, K.J. & Fendorf, S.: Contrasting effects of dissimilatory iron(III) and arsenic(V) reduction on arsenic retention and transport, *Environ. Sci. Technol.* 40 (2006), pp. 6715–6721.

Kohler, M., Hofmann, K., Volsgen, F., Thurow, K. & Koch, A.: Bacterial release of arsenic ions and organoarsenic compounds from soil contaminated by chemical warfare agents. *Chemosphere* 42 (2001), pp. 425–429.

Krumova, K., Nikolvska, M. & Groudeva, V.: Isolation and identification of arsenic-transforming bacteria from arsenic contaminated sites in Bulgaria. *Biotechnol.* 22:2 (2008), pp. 721–728.

Kumar, G.P., Yadav, S.K., Thawale, P.R., Singh, S.K. and Juwarkar, A.A.: Growth of *Jatropha curcas* on heavy metal contaminated soil amended with industrial wastes and *Azotobacter* – A greenhouse study. *Bioresour. Technol.* 99 (2008), pp. 2078–2082.

Langley, S. & Beveridge, T.J.: Effect of O-side-chain-lipopolysaccharide chemistry on metal binding. *Appl. Environ. Microbiol.* 65 (1999), pp. 489–498.

Ledin, M.: Accumulation of metals by microorganisms-processes and importance for soil systems. *Earth-Sci. Rev.* 51 (2000), pp. 1–31.

Lee, K.Y., Yoon, I.H., Lee, B.T., Kim, S.O. & Kim, K.W.: A novel combination of anaerobic bioleaching and electrokinetics for arsenic removal from mine tailing soil. *Environ. Sci. Technol.* 43 (2009), pp. 9354–9360.

Lee, S.J., Lee, S.C., Choi, S.H., Chung, M.K., Rhie, H.G. & Lee, S.H.: Effect of ArsA, arsenite-specific ATPase, on inhibition of cell division in *Escherichia coli. J. Microbiol. Biotechnol.* 11 (2001), pp. 825–830.

Lehr, C.R., Polishchuk, E., Delisle, M.C., Franz, C. & Cullen, W.R.: Arsenic methylation by micro-organisms isolated from sheepskin bedding materials. *Hum. Exp. Toxicol.* 22 (2003), pp. 325–334.

Leung, H.M., Ye, Z.H. & Wong, M.H.: Interactions of mycorrhizal fungi with *Pteris vittata* (As hyperaccumulator) in As-contaminated soils. *Environ. Pollut.* 139 (2006), pp. 1–8.

Leung, K., Cassidy, M.B., Holmes, S.B., Lee, H. & Trevors, J.T.: Survival of k-carrageenan-encapsulated and unencapsulatad *Pseudomones aeruginosa* UG2Lr cells in forest soil monitored by polymerese chain reaction and spread plating. *FEMS Microbio/Ecol* 16 (1995), pp. 71–82.

Leyval, C., Turnau, K. & Haselwandter, K.: Interactions between heavy metals and mycorrhizal fungi in polluted soils: physiological, ecological and applied aspects. *Mycorrhiza* 7 (1997), pp. 139–153.

Li, H., Ye, Z.H., Chan, W.F., Chen, X.W., Wu, F.Y., Wu, S.C. & Wong, M.H.: Can arbuscular mycorrhizal fungi improve grain yield, As uptake and tolerance of rice grown under aerobic conditions? *Environ. Pollut.* 159 (2011), pp. 2537–2545.

Liu, S., Zhang, F., Chen, J. & Sun, G.X.: Arsenic removal from contaminated soil via biovolatilization by genetically engineered bacteria under laboratory conditions. *J. Environ. Sci.* 23 (2011), pp. 1544–1550.

Liu, W.J., Zhu, Y.G., Smith, F.A. & Smith, S.E.: Do iron plaque and genotypes affect arsenate uptake and translocation by rice seedlings (*Oryza sativa* L.) grown in solution culture? *J. Experim. Bot.* 55 (2004), pp. 1707–1713.

Liu, Y., Zhu, Y.G., Chen, B.D., Christie, P. & Li, X.L.: Influence of the arbuscular mycorrhizal fungus *Glomus mosseae* on uptake of arsenate by the As hyperaccumulator fern *Pteris vittata* L. *Mycorrhiza* 15 (2005a), pp. 187–192.

Liu, Y., Zhu, Y.G., Chen, B.D., Christie, P. & Li, X.L.: Yield and arsenate uptake of arbuscular mycorrhizal tomato colonized by *Glomus mosseae* BEG167 in As spiked soil under glasshouse conditions. *Environ. Int.* 31 (2005b), pp. 867–873.

Liu, Y., Christie, P., Zhang, J. & Li, X.: Growth and arsenic uptake by Chinese brake fern inoculated with an arbuscular mycorrhizal fungus. *Environ. Experim. Bot.* 66 (2009), pp. 435–441.

Lloyd, J.R. & Oremland, R.S.: Microbial transformation of arsenic in the environment from soda lake to aquifers. *Elements* 2 (2006), pp. 85–90.

Lorenzo, V. de.: Systems biology approaches to bioremediation. *Curr. Opinion Biotechnol.* 19:6 (2008), pp. 579–589.

Lovley, D.R. & Coates, J.D.: Bioremediation of metal contamination. *Curr. Opinion Biotechol.* 8 (1997), pp. 285–287.

Lovley, D.R., Giovannoni, S.J., White, D.C., Champine, J.E., Phillips, E.J.P., Gorby, Y.A. & Goodwin, S.: *Geobacter metallireducens* gen. nov. sp. Nov., a microorganism capable of coupling the complete oxidation of organic compounds to the reduction of iron and other metals. *Arch. Microbiol.* 159 (1993), pp. 336–344.

Lu, F.J.: Blackfoot disease: arsenic or humic acid. *Lancet* 336 (1990), pp. 115–116.

Ma, J.F., Yamaji, N., Mitani, N., Xu, X.Y., Su, Y.H., McGrath, S.P. & Zhao, F.J.: Transporters of arsenite in rice and their role in arsenic accumulation in rice grain. *PNAS* 105 (2008), pp. 9931–9935.

Ma, L.Q., Komar, K.M., Tu, C., Zhang, W., Cai, Y. & Kennelley, E.D.: A fern that hyperaccumulates arsenic. *Nature* 409 (2001) p. 579.

Macur, R.E., Jackson, C.R., Botero, L.M., McDermott, T.R. & Inskepp, W.P.: Bacterial populations associated with the oxidation and reduction of arsenic in an unsaturated soil. *Environ. Sci. Technol.* 38 (2004), pp. 104–111.

Macur, R.E., Wheeler, J.T. & McDermott, T.R.: Microbial populations associated with the reduction and enhanced mobilization of arsenic in mine tailings. *Environ. Sci. Technol.* 35(18) (2001), pp. 3676–3682.

Macy, J.M., Santini, J.M., Panling, B.V., O'Neil A.H. & Sly, L.I.: Two new arsenate/sulfate-reducing bacteria: Mechanisms of arsenate reduction. *Arch. Microbiol.* 173 (2000), pp. 49–57.

Maheswari, S. & Murugesan, A.G.: Remediation of arsenic in soil by *Aspergillus nidulans* isolated from an arsenic-contaminated site. *Environ. Technol.* 30:9 (2009), pp. 921–926.

Malajczuk, N. & Cromack, K.: Accumulation of calcium oxalate in the mantle of ectomycorrhizal roots of *Pinus radiate* and *Eucalyptus marginata*. *New Phytol.* 92 (1982) pp. 527–531.

Marschner, H.: *Mineral nutrition of higher plants*. 2nd ed., Academic Press, London, UK, 1995.

Mateos, L.M., Ordonez, E., Letek, M. & Gil, J.A.: *Corynebacterium glutamicum* as a model bacterium for bioremediation of arsenic. *Int. Microbiol.* 9 (2006), pp. 207–215.

McBride & Wolfe, R.S.: Biosynthesis of dimethylasrine by a methanobacterium, *Biochemistry* 10 (1971), pp. 4312–4317.

McLaren, S.J. & Kim, N.D.: Evidence for a seasonal fluctuation of arsenic in New Zealand's longest river and the effect of treatment on concentrations in drinking water. *Environ. Pollut.* 90 (1995), pp. 67–73.

McLoughlin, A.J.: Controlled release of immobilized cells as a strategy to regulate ecological competence of inocula. In: T. Scheper (ed): *Biotechnics/wastewater*. Springer, Berlin, Germany, 1994, pp. 1–45.

Meharg, A.A. & Hartley-Whitaker, J.: Arsenic uptake and metabolism in arsenic resistant and nonresistant plant species. *New Phytol.* 154 (2002), pp. 29–43.

Meharg, A.A. & Rahman, M.M.: Arsenic contamination of Bangladesh paddy field soils: implications for rice contribution to arsenic consumption. *Environ. Sci. Technol.* 37 (2003), pp. 229–234.

Meng, X.-Y., Qin, J., Wang, L.H., Duan, G.L., Sun, G.X., Wu, H.L., Chu, C.C., Ling, H.Q., Rosen, B.P. & Zhu, Y.G.: Arsenic biotransformation and volatilization in transgenic rice. *New Phytol.* 19:1 (2011), pp. 49–56.

Messens, J., Martins, J.C., Brosens, E., Van Belle, K., Jacobs, D.M., Willem, R. & Wyns, L.: Kinetics and active site dynamics of *Staphylococcus aureus* arsenate reductase. *J. Biol. Inorg. Chem.* 7 (2002), pp. 146–156.

Michalke, K., Wickenheiser, E.B., Mehring, M., Hirner A.V. & Hensel R.: Production of volatile derivatives of metal(loid)s by microflora involved in anaerobic digestion of sewage sludge. *Appl. Environ. Microbiol.* 66 (2000), pp. 2791–2796.

Mobley, H.L.T., & Rosen, B.P.: Energetics of plasmidmediated arsenate resistance in *Escherichia coli. PNAS* 79 (1982), pp. 6119–6122.

Mougel, C., Offre, P., Ranjard, L., Corberand, T., Gamalero, E., Robin, C. & Lemanceau, P.: Dynamic of the genetic structure of bacterial and fungal communities at different developmental stages of *Medicago truncatula* Gaertn. cv. *Jemalong line J5. New Phytol.* 170 (2006), pp. 165–175.

Mukherjee, A., Das, D., Kumar Mandal, S., Biswas, R., Kumar Das, T., Boujedani, N. & Khuda-Bakhsh, A.R.: Tolerance of arsenate-induced stress in *Aspergillus niger*, a possible candidate for bioremediation. *Ecotoxicol. Environ. Saf.* 73 (2010), pp. 172–182.

Mukhopadhyay, R., Shi, J. & Rosen, B.P.: Purification and characterization of Acr2p, the *Saccharomyces cerevisiae* arsenate reductase. *J. Biol. Chem.* 275 (2000), pp. 21,149–21,157.

Murugesan, G.S., Sathishkumar, M. & Swaminathan, K.: Arsenic removal from groundwater by pretreated waste tea fungal biomass. *Bioresour. Technol.* 97:3 (2006), pp. 483–487.

Nagvenkar, G.S. & Ramaiah, N.: Arsenite tolerant and biotransformation potential in estuarine complex. *Ecotoxicol.* 19:4 (2010), pp. 604–613.

Newman, D.K., Kennedy, E.K., Coates, J., Ahmann, D., Ellis, D., Lovely, D. & Morel, F.: Dissimilatory arsenate and sulfate reduction in *Desulfotomaculum auripigmentum* sp. *Arch. Microbiol.* 168 (1997), pp. 380–388.

Newman, D.K., Ahmann, D. & Morel, F.M.M.: A brief review of microbial arsenate respiration. *Geomicrobiol. J.* 15 (1998), pp. 255–268.

Nie, L., Shah, S., Rashid, A., Burd, G.I., Dixon, D.G. & Glick, B.R.: Phytoremediation of arsenate contaminated soil by transgenic canola and the plant growth-promoting bacterium *Enterobacter cloacae* CAL2. *Plant Physiol. Biochem.* 40 (2002), pp. 355–361.

Niggemeyer, A., Spring, S., Stackenbrandt, E. & Rosenzweig, R.F.: Isolation and characterization of a novel As(V)-reducing bacterium: Implication for arsenic mobilization and the genus *Desulfitobacterium*. *Appl. Environ. Microbiol.* 67 (2001), pp. 5568–5580.

NRC (National Research Council): *Arsenic in drinking water*. National Academy Press, Washington, DC, 1999.

Nunan, N., Daniell, T.J., Singh, B.K., Papert, A., McNicol, J.W. & Prosser, J.I.: Links between plant and rhizoplane bacterial communities in grassland soils, characterized using molecular techniques. *Appl. Environ. Microbiol.* 71 (2005), pp. 6784–6792.

Ongley, L.K., Sherman, L., Armienta, A., Concilio, A. & Salinas, C.F.: Arsenic in the soils of Zimapan, Mexico. *Environ. Pollut.* 145 (2007), pp. 793–799.

Ordonez, F., Letek, M., Valbuena, N., Gil, J.A. & Mateos, L.M.: Analysis of genes involved in arsenic resistance in *Corynebacterium glutamicum* ATCC 13032. *Appl. Environ. Microbiol.* 71 (2005), pp. 6206–6215.

Ordonez, E., Van Belle, K., Roos, G., Galan, S. De, Letek, M., Gil, J.A., Wyns, L., Mateos, L.M. & Messens, J.: Arsenate reductase, mycothiol, and mycoredoxin concert thiol/disulfide exchange. *The J. Biol. Chemi.* 284:22 (2009), pp. 15,107–15,116.

Oremland, R.S. & Stolz, J.F.: Dissimilatory reduction of selenate and arsenate in nature. In D.R. Lovley (ed): *Environmental Metal-Microbe Interaction*. Amer. Soc. Microbiology Press, Washington, DC, 2000, pp. 199–224.

Oremland, R.S. & Stolz, J.F.: Dissimilatory reduction of selenate and arsenate in nature. In: D.R. Lovely (ed): *Environmental microbe-metal interactions*. American Society for Microbiology Press, Washington, DC, 2001, pp. 199–224.

Oremland, R.S. & Stolz, J.F.: The ecology of arsenic. *Science* 300 (2003), pp. 939–944.

Oremland, R.S. & Stolz, J.F.: Arsenic, microbes and contaminated aquifers. *Trends Microbiol.* 13 (2005), pp. 45–49.

Oremland, R.S., Newman, D.K., Kail, B.W. & Stolz, J.F.: Bacterial respiration of arsenate and its significance in the environment. In: W.T. Frankenberger Jr (ed): *Environmental chemistry of arsenic*. Marcel Dekker, Inc., New York, NY, 2002, pp. 273–296.

Owsianiak, M., Dechesne, A., Binning, P.J., Chambon, J.C., Sørensen, S.R. & Smets, B.F.: Evaluation of bioaugmentation with entrapped degrading cells as a soil remediation technology. *Environ. Sci. Technol.* 44 (2010), pp. 7622–7627.

Paez-Espino, D., Tamames, J., de Lorenzo, V. & Cánovas, D.: Microbial responses to environmental arsenic. *Biometals* 22 (2009), pp. 117–130.

Page, A.L., Elseewi, A.A. & Straughan, I.R.: Physical and chemical properties of fly ash from coal fired power plants. *Res. Review* 71 (1979), pp. 83–120.

Pajuelo, E., Rodríguez-Llorente, I.D., Dary, M. & Palomares, A.J.: Toxic effects of arsenic on *Sinorhizobium–Medicago sativa* symbiotic interaction. *Environ. Pollut.* 154 (2008), pp. 203–211.

Panigrahi, D.P. & Randhawa, G.S.: A novel method to alleviate arsenic toxicity in *alfalfa* plants using a deletion mutant strain of *Sinorhizobium meliloti*. *Plant Soil* 336 (2010), pp. 459–467.

Patel, P.C., Goulhen, F., Boothman, C., Gault, A.G., Charnock, J.M., Kalia, K. & Lloyd J.R.: Arsenate detoxification in a *Pseudomonas* hypertolerant to arsenic. *Arch. Microbiol.* 187 (2007), pp. 171–183.

Pearce, F.: Arsenic in the water. *The Guardian* (London), Online section (19 February) (1998), pp. 2–3.

Pepi, M., Volterrani, M., Renzi, M., Marvasi, M., Gasperini, S., Franchi, E. & Focardi, S.E.: Arsenic-resistant bacteria isolated from contaminated sediment of the Orbetello Lagoon, Italy, and their characterization. *Appl. Microbiol.* 103 (2007), pp. 2299–2308.

Pillichshammer, M., Pumpel, T., Poder, R., Eller, K., Klima, J. & Schinner, F.: Biosorption of chromium to fungi. *Biometals* 8 (1995), pp. 117–121.

Pokhrel, D. & Viraraghavan, T.: Arsenic removal from an aqueous solution by a modified fungal biomass. *Water Res.* 40 (2006), pp. 549–552.

Polemio, M., Senesi, S. & Bufo, S.A.: Soil contamination by metal – a survey in industrial and rural areas of southern Italy. *Sci. Total Environ.* 25 (1982), pp. 71–79.

Qin, J., Rosen, B.P., Zhang, Y., Wang, G.J., Franke, S. & Rensing, C.: Arsenic detoxification and evolution of trimethylarsine gas by a microbial arsenite S-adenosylmethionine methyltransferase. *PNAS* 103 (2006) pp. 2075–2080.

Qin, J., Lehr, C.R., Yuan, C., Le, X.C., Dermott, T.R.M.C. & Prosen, B.: Biotransformation of arsenic by a Yellowstone thermoacidophilic eukaryotic alga. *PNAS* 106:13 (2009), pp. 5213–5217.

Quéméneur, M., Heinrich-Salmeron, A., Muller, D., Lièvremont, D., Jauzein, M., Bertin, P.N., Garrido, F. & Joulian, C.: Diversity surveys and evolutionary relationships of *aoxB* genes in aerobic arsenite-oxidizing bacteria. *Appl. Environ. Microbiol.* 74:14 (2008), pp. 4567–4573.

Rajkumar, M., Ae, N., Prasad, M.N.V. & Freitas, H.: Potential of siderophore-producing bacteria for improving heavy metal phytoextraction. *Trends Biotechnol.* 28 (2010), pp. 142–149.

Reed, M.L.E. & Glick, B.R.: Growth of canola (*Brassica napus*) in the presence of plant growth-promoting bacteria and either copper or polycyclic aromatic hydrocarbons. *Canad. J. Microbiol.* 51: (2005), pp. 1061–1069.

Reichman, S.M.: The potential use of the legume *rhizobium* symbiosis for the remediation of arsenic contaminated sites. *Soil Biol. Biochem.* 39 (2007), pp. 2587–2593.

Reinhart, K.O. & Callaway, R.M.: Soil biota and invasive plants. *New Phytol.* 170 (2006), pp. 445–457.

Rhine, E.D., Phelps, C.D. & Young, L.Y.: Anaerobic arsenite oxidation by novel denitrifying isolates. *Environ. Microbiol.* 8 (2006), pp. 899–908.

Rochelle, P.A., Wetherbee, M.K. & Olson, B.H.: Distribution of DNA sequences encoding narrow- and broad-spectrum mercury resistance. *Appl. Environ. Microbiol.* 57 (1991), pp. 1581–1589.

Romero, F.M., Armienta, M.A. & Carrillo, C.A.: Arsenic sorption by carbonate-rich aquifer material, a control on arsenic mobility at Zimapan. Mexico. *Arch. Environ. Contam. Toxicol.* 47 (2004), pp. 1–13.

Rosen, B.P.: Families of arsenic transporters. *Trends Microbiol.* 7 (1999), pp. 207–212.

Rosen, B.P.: Biochemistry of arsenic detoxification. *FEBS Letters* 529 (2002), pp. 86–92.

Ross, S.M. & Kae, K.J.: The meaning of metal toxicity in soil-plant systems. In: S.M. Ross (ed): *Toxic metals in soil-plant systems*. John Wiley and Sons, New York, NY, 1994, pp. 153–188.

Ross, A.D., Lawrie, R.A., Whatmuff, M.S., Keneally, J.P. & Awad, A.S.: Guidelines for the use of sewage sludge on agricultural land. New South Wales Agriculture, Sydney, Australia, 1991.

Roussel, C., Bril, H. & Fernandez, A.: Arsenic speciation: involvement in evaluation of environmental impact caused by mine wastes. *J. Environ. Qual.* 29 (2000), pp. 182–188.

Rowland, H.A.L., Pederick, R.L., Polya, D.A., Pancoast, R.D., Van Dongen, B.E., Gault, A.G., Vaugh, D.J., Bryant, C., Anderson, B. & Lloyd, J.R.: The control of organic matter on microbially mediated iron reduction and arsenic release in shallow alluvial aquifers, Cambodia. *Geobiology* 5 (2007), pp. 281–292.

Roychowdhury, T., Uchino, T., Tokunaga, H. & Ando, M.: Survey of arsenic in food composites from an arsenic affected area of West Bengal, India. *Food Chem. Toxicol.* 40 (2002), pp. 1611–1621.

Russeva, E.: Speciation analysis – pecularities and requirements. *Anal. Lab.* 4 (1995), pp. 143–148.

Sadiq, M.: Arsenic chemistry in soils: An overview of thermodynamic predictions and field observations. *Water Air Soil Pollut.* 93 (1997), pp. 117–136.

Saltikov, C.V. & Olson, B.H.: Homology of *Escherichia coli* R773 arsA, arsB and arsC genes in arsenic-resistant bacteria isolated from raw sewage and arsenic enriched creek waters. *Appl. Environ. Microbiol.* 68 (2002), pp. 280–288.

Santini, J.M. & Vanden Hoven, R.N.: Molybdenum-containing arsenite oxidase of the chemolithoautotrophic arsenite oxidizer NT-26. *J. Bacteriol.* 186 (2004), pp. 1614–1619.

Santini, J.M., Sly, L.I., Schnagl, R.D. & Macy J.M.: A new chemolithoautotrophic arsenite-oxidizing bacterium isolated from a gold mine: Phylogenetic, physiological, and preliminary biochemical studies. *Appl. Environ. Microbiol.* 66 (2000), pp. 92–97.

Sauge-Merle, S., Cuine, S., Carrier, P., Lecomte-Pradines, C., Luu, D.T. & Peltier, G.: Enhanced toxic metal accumulation in engineered bacterial cells expressing *Arabidopsis thaliana* phytochelatinsynthase. *Appl. Environ. Microbiol.* 69 (2003), pp. 490–494.

Say, R., Yilmaz, N. & Denizli, A.: Removal of heavy metalions using the fungus *Penicillium canescens*. *Adsorpt. Sci. Technol.* 21 (2003), pp. 643–650.

Schiewer, S., & Volesky, B.: Biosorption processes for heavy metal removal. In: D.R. Lovley (ed): *Environmental microbe–metal interactions*. ASM Press, Washington, DC, 2000, pp. 329–362.

Schoof, R.A., Yost, L.J., Eickhoff, J., Crecelius, E.A., Meacher, D.M. & Menzel, D.B.: A market basket survey of inorganic arsenic in food. *Food Chem. Toxicol.* 37 (1999), pp. 839–836.

Schue, M., Dover, L.G., Besra, G.S., Parkhill, J. & Brown, N.L.: Sequence and analysis of a plasmid encoded mercury resistance operon from *Mycobacterium marinum* identifies MerH, a new mercuric ion transporter. *J. Bacteriol.* 19 (2009), pp. 439–444.

Seidel, H., Mattusch, J., Wennrich, R., Morgenstern, P. & Ondruschka, J.: Mobilization of arsenic and heavy metals from contaminated sediments by changing the environmental conditions. *Acta Biotechnol.* 22 (2002), pp. 153–160.

Shariatpanahi, M., Anderson, A.C., Abdelghani, A.A., Englande, A.J., Hughes, J. & Wilkinson, R.F.: Biotransformation of the pesticide sodium arsenate. *J. Environ. Sci. Health* B 16:1 (1981), pp. 35–47.

Sharples, J.M., Meharg, A.A., Chambers, S.M. & Cairney, J.W.: Mechanism of arsenate resistance in the ericoid mycorrhizal fungus *Hymenoscyphus ericae*. *Plant Physiol.* 124 (2000), pp. 1327–1334.

Silver, S. & Keach, D.: Energy-dependent arsenate efflux: The mechanism of plasmid-mediated resistance. *PNAS* 79 (1982), pp. 6114–6118.

Silver, S. & Phung, L.T.: Bacterial heavy metal resistance: New surprises. *Annu. Rev. Microbiol.* 50 (1996), pp. 753–789.

Silver, S. & Phung, L.T.: Genes and enzymes involved in bacterial oxidation and reduction of inorganic arsenic. *Appl. Environ. Microbiol.* 71 (2005), pp. 599–608.

Simeonova, D.D., Lievremont, D., Lagarde, F., Muller, D.A., Groudeva, V.I. & Lett, M.C.: Microplate screening assay for the detection of arsenite oxidizing and arsenate reducing bacteria. *FEMS Microbiol. Lett.* 237:2 (2004), pp. 249–253.

Singh, P. & Cameotra, S.S.: Enhancement of metal bioremediation by use of microbial surfactants. *Biochem. Biophys. Res. Com.* 319 (2004), pp. 291–297.

Smedley, P.L. & Kinniburgh, D.G.: A review of the source, behavior and distribution of arsenic in natural waters. *Appl. Geochem.* 17 (2002), pp. 517–568.

Smith, A., Lingas, E. & Rahman, M.: Contamination of drinking-water by arsenic in Bangladesh: a public health emergency. *Bull. World Health Organ.* 78 (2000), pp. 1093–1103.

Smith, S.L. & Jaffe, P.R.: Modeling the transport and reaction of trace metals in water-saturated soils and sediments. *Water Resour. Res.* 34 (1998), pp. 3135–3147.

Soeroes, C., Kienzl, N., Ipolyi, I., Dernovics, M., Fodor, P. & Kuehnelt, D.: Arsenic uptake and arsenic compounds in cultivated *Agaricus bisporus*. *Food Control* 16:5 (2005), pp. 459–464.

Srivastava, D., Madamwar, D. & Subramanian, R.B.: Pentavalent arsenate reductase activity in cytosolic fractions of *Pseudomonas* sp., isolated from arsenic-contaminated sites of Tezpur, Assam. *Appl. Biochem. Biotechnol.* 162 (2010), pp. 766–779.

Srivastava, P.K., Vaish, A., Dwivedi, S., Chakrabarty, D., Singh, N. & Tripathi, R.D.: Biological removal of arsenic pollution by soil fungi. *Sci. Total Environ.* 409 (2011), pp. 2430–2442.

Stilwell, D.E. & Gorny, K.D.: Contamination of soil with copper chromium and arsenic under decks built with pressure treated wood. *Bull. Environ. Contam. Toxicol.* 58 (1997), pp. 22–29.

Stolz, J.F. & Oremland, R.S.: Bacterial respiration of arsenic and selenium. *FEMS Microbiol. Rev.* 23 (1999), pp. 615–627.

Stolz, J.F., Basu, P. & Oremland, R.S.: Microbial arsenic metabolism: new twists on an old poison. *Microbe* 5 (2010), pp. 53–59.

Su, S., Zeng, X., Bai L., Jiang X. & Li L.: Bioaccumulation and biovolatalization of pentavalent arsenic by *Penicillin janthinellum, Fusarium oxysporum, Trichoderma asperellum* under laboratory conditions. *Curr. Microbiol.* 61 (2010a), pp. 261–266.

Su, Y.H., McGrath, S.P. & Zhao, F.J.: Rice is more efficient in arsenite uptake and translocation than wheat and barley. *Plant Soil* 328 (2010b), pp. 27–34.

Takeuchi, M., Kawahata, H., Gupta, L.P., Kita, N., Morishita, Y. & Komai T.: Arsenic resistance and removal by marine and non-marine bacteria. *J. Biotechnol.* 127 (2007), pp. 434–442.

Tamaki, S. & Frankenberger, Jr W.T.: Environmental biochemistry of arsenic. *Rev. Environ. Contam. Toxicol.* 124 (1992), pp. 79–110.

Tang, S.R., Wilke, B. & Brooks, R.R.: Heavy-metal uptake by metal-tolerant *Elsholtzia aplendens* and *Commelina communis* from China. *Commun. Soil Sci. Plant Anal.* 32:56 (2001), pp. 895–905.

Tauriainen, S., Karp, M., Chang, W. & Virta, M.: Recombinant luminescent bacteria for measuring bioavailable arsenite and antimonite. *Appl. Environ. Microbiol.* 63 (1997), pp. 4456–4461.

Tereshina, V.M., Marin, A.P., Kosyakov, V.N., Kozlov, V.P. & Feofilova, E.P.: Different metal sorption capacities of cell wall polysaccharides of *Aspergillus niger*. *Appl. Biochem. Microbiol.* 35 (1999), pp. 389–392.

Thayer, J.S.: *Organometallic compounds and living organisms*. Academic Press, Orlando, FL, 1984.

Thimmaraju, R., Czymmek, K.J., Pare, P.W. & Bais, H.P.: Root-secreted malic acid recruits beneficial soil bacteria. *Plant Physiol.* 148 (2008), pp. 1547–1556.

Thomson, B., Aragon, A., Anderson, J., Chwirka, J. & Brady, P.: Rapid small scale column testing for evaluating arsenic adsorbents. American Water Works Association (AWWA), 2005.

Trotta, A., Falaschi, P., Cornara, L., Minganti, V., Fusconi, A., Drava, G. & Berta, G.: Arbuscular mycorrhizae increase the arsenic translocation factor in the As hyperaccumulating fern *Pteris vittata* L. *Chemosphere* 65 (2006), pp. 74–81.

Tufano, K.J., Reyes, C., Saltikov, C.W. & Fendorf, S.: Reductive processes controlling arsenic retention: Revealing the relative importance of iron and arsenic reduction. *Environ. Sci. Technol.* 42 (2008a), pp. 8283–8289.

Tufano, K.T. & Fendorf, S.: Contrasting impacts of iron reduction on arsenic retention. *Environ. Sci. Technol.* 42 (2008b), pp. 4777–4783.

Turpeinen, R., Pantsar-Kallio, M., Haggblom, M. & Kairesalo, T.: Influence of microbes on mobilization, toxicity and biomethylation of arsenic in soil. *Sci. Total Environ.* 236 (1999), pp. 173–180.

Turpeinen, R., Salminen, J. & Kairesalo, T.: Mobility and bioavailability of lead in contaminated boreal forest soil. *Environ. Sci. Technol.* 34 (2000), pp. 5152–5156.

Turpeinen, R., Pantsar-Kallio, M. & Kairesalo, T.: Role of microbes in controlling the speciation of arsenic and production of arsines in contaminated soils. *Sci. Total Environ.* 285 (2002), pp. 133–145.

Turpeinen, R., Kairesalo, T. & Häggblom, M.M.: Microbial community structure and activity in arsenic-, chromium- and copper-contaminated soils. *FEMS Microbiol. Ecolol.* 47 (2004), pp. 39–50.

Tyagi, M., Manuela, M., da Fonseca, R. & de Carvalho, C.C.C.R.: Bioaugmentation and biostimulation strategies to improve the effectiveness of bioremediation processes. *Biodegrad.* 22 (2011), pp. 231–241.

Vala, A.K., Davariya, V. & Upadhyay, R.V.: An investigation on tolerance and accumulation of a facultative marine fungus *Aspergillus flavus* to pentavalent arsenic. *J. Ocean Univ. China* (*Oceanic and Coastal Sea Research*) 9: (2010), pp. 65–67.

Valls, M., Atrian, S., de Lorenzo, V. & Fernandez, L.A.: Engineering a mouse metallothionein on the cell surface of *Ralstonia eutropha* CH34 for immobilization of heavy metals in soil. *Nat. Biotechnol.* 18 (2000), pp. 661–665.

Valls, M. & De Lorenzo, V.: Exploiting the genetic and biochemical capacities of bacteria for the remediation of heavy metal pollution. *FEMS Microbiol. Rev.* 26 (2002), pp. 327–338.

van der Gast, C.J., Whiteley, A.S. & Thompson, I.P.: Temporal dynamics and degradation activity of an bacterial inoculums for treating waste metal-working fluid. *Environ. Microbiol.* 6 (2004), pp. 254–263.

Visoottiviseth, P. & Panviroj, P.: Selection of fungi capable of removing toxic arsenic compounds from liquid medium. *Science Asia* 27 (2001), pp. 83–92.

Vogel-Mikus, K., Pongrac, P., Kump, P., Necemer, M. & Regvar, M.: Colonisation of a Zn, Cd and Pb hyperaccumulator *Thlaspi praecox* Wulfen with indigenous arbuscular mycorrhizal fungal mixture induces changes in heavy metal and nutrient uptake. *Environ. Pollut.* 139 (2006), pp. 362–371.

Volesky, B. & Holan, Z.R.: Biosorption of heavy metals. *Biotechnol. Prog.* 11 (1995), pp. 235–250.

Walsh, L.M., Sumner, M.E. & Keeney, D.R.: Occurrence and distribution of arsenic in soils and plants. *Environ. Health Perspect.* 19 (1977), pp. 67–71.

Wang, C. & Lazarides, E.: Arsenite-induced changes in methylation of the 70,000 Dalton heat shock proteins in chicken embryo fibroblasts. *Biochem. Biophys. Res. Commun.* 119 (1984), pp. 735–743.

Wang, F.Y., Lin, X.G. & Yin, R.: Inoculation with arbuscular mycorrhizal fungus *Acaulospora mellea* decreases Cu phytoextraction by maize from Cu-contaminated soil. *Pedobiologia* 51:2 (2007), pp. 99–109.

Wang, Q., Xiong, D., Zhao, P., Yu, X., Tu, B. & Wang, G.: Effect of applying an arsenic-resistant and plant growth–promoting *rhizobacterium* to enhance soil arsenic phytoremediation by *Populus deltoides* LH05-17. *J. Appl. Microbiol.* 111:5 (2011), pp. 1065–1074

Wang, S. & Zhao, X.: On the potential of biological treatment for arsenic contaminated soils and groundwater. *J. Environ. Manag.* 90 (2009), pp. 2367–2376.

Watt, M., Silk, W.K. & Passioura, J.B.: Rates of root and organism growth, soil conditions, and temporal and spatial development of the rhizosphere. *Ann. Bot.* 97 (2006), pp. 839–855.

Weeger, W., Lievremont. D., Perret, M., Lagarde, F., Hubert, J.C., Leroy, M. & Lett, M.C.: Oxidation of arsenite to arsenate by a bacterium isolated from an aquatic environment. *Biometals* 12:2 (1999), pp. 141–149.

Wei, H.X.U., Huai, L., Mac, Q.F. & Xiongd, Z.T.: Root exudates, rhizosphere Zn fractions, and Zn accumulation of ryegrass at different soil Zn levels. *Pedosphere* 17:3 (2007), pp. 389–396.

Weiss, J.V., Megonigal, J.P., Emerson, D. & Backer, S.M.: Enumeration of Fe(II)-oxidizing and Fe(III)-reducing bacteria in the root-zone of wetland plants: Implications for a rhizosphere Fe cycle. *Biogeochemistry* 64 (2003), pp. 77–96.

WHO: Guidelines for drinking-water quality – Volume 1: Recommendations. Third edition, incorporating first and second addenda World Health Organization, Geneva, Switzerland, 2008

Williams, P.N., Price, A.H., Raab, A., Hossain, S.A., Feldman, J. & Meharg, A.A.: Variation in arsenic speciation and concentration in paddy rice related to dietary exposure. *Environ. Sci. Technol.* 39 (2005), pp. 5531–5540.

Williams, P.N., Islam, M.R., Adomako, E.E., Raab, A., Hossain, S.A., Zhu, Y.G. & Meharg, A.A.: Increase in rice grain arsenic for regions of Bangladesh irrigating paddies with elevated arsenic in groundwater. *Environ. Sci. Technol.* 40 (2006), pp. 4903–4908.

Williams, P.N., Villada, A., Deacon, C., Raab, A., Figuerola, J., Green, A.J., Feldmann, J. & Meharg, A.A.: Greatly enhanced arsenic shoot assimilation in rice leads to elevated grain levels compared to wheat and barley. *Environ. Sci. Technol.* 41 (2007), pp. 6854–6859.

Wood, C.R.: Water quality of large discharges from mines in the anthracite region of eastern Pennsylvania. USGS Water Resources Investigation Report 95-4243, 1996.

Woolson, E.A. & Kearney, P.C.: Persistence and reactions of 14C-cacodylic acid in soils. *Environ. Sci.Technol.* 7 (1973), pp. 47–50.

Wu, F.Y., Ye, Z.H. & Wong, M.H.: Intra specific differences of arbuscular mycorrhizal fungi in their impacts on arsenic accumulation by *Pteris vittata* L. *Chemosphere* 76 (2009), pp. 1258–1264.

Wu, J., Tisa, L.S. & Rosen B.P.: Membrane topology of the ArsB protein, the membrane subunit of an anion-translocating ATPase. *J. Biol. Chem.* 267 (1992), pp. 12570–12576.

Wysocki, R. & Tamas, M.J.: How *Saccharomyces cerevisiae* copes with toxic metals and metalloids. *FEMS Microbiol. Rev.* 34 (2010), pp. 925–951.

Xu, X.Y., McGrath, S.P., Meharg, A.A. & Zhao, F.J.: Growing rice aerobically markedly decreases As accumulation. *Environ. Sci. Technol.* 42 (2008), pp. 5574–5579.

Yamamura, S., Yamashita, M., Fujimoto, M., Kuroda, N., Kashiwa, M., Sei, M., Fujita, K. & Ike, M.: *Bacillus selenatarsenatis* sp.nov., a selenate- and arsenate-reducing bacterium isolated from the effluent drain of a glass-manufacturing plant. *Int. J. Syst. Evol. Microbiol.* 57 (2007) pp. 1060–1064.

Yan, G. & Viraraghavan, T.: Heavy-metal removal from aqueous solution by fungus *Mucor rouxii*. *Water Res,* 37 (2003), pp. 4486–4496.

Yang, Q., Tu, S., Wang, G., Liao, X. & Yan, X.: Effectiveness of applying of arsenate reducing bacteria to enhance arsenic removal from polluted soils by *Pteris vittata* L. *Int. J. Phytorem.* 14 (2011), pp. 89–99.

Yoshinaga, M., Cai, Y. & Rosen, B.P.: Demethylation of methylarsonic acid by a microbial community. *Environ. Microbiol.* 13:5 (2011), pp. 1205–1215.

Yu, Y., Zhang, S., Huang, H., Luo, L. & Wen, B.: Arsenic accumulation and speciation in *Maize* as affected by inoculation with arbuscular mycorrhizal Fungus *Glomus mosseae*. *J. Agric. Food Chem.* 57 (2009), pp. 3695–3701.

Yuan, C., Lu, X., Qin, J., Rosen, B.P. & Le, X.C.: Volatile arsenic species released from *Escherichia coli* expressing the AsIII s-adenosylmethionine methyltransferase gene. *Environ. Sci. Technol.* 42:9 (2008), pp. 3201–3206.

Zafar, S., Aqil, F. & Ahmad, I.: Metal tolerance and biosorption potential of filamentous fungi isolated from metal contaminated agricultural soil. *Bioresour. Technol.* 98:13 (2007), pp. 2557–2561.

Zaidi, S., Usmani, S., Singh, B.R. & Musarrat, J.: Significance of *Bacillus subtilis* strain SJ-101 as a bioinoculant for concurrent plant growth promotion and nickel accumulation in *Brassica juncea*. *Chemosphere* 64 (2006), pp. 991–997.

Zavala, Y.J., Gerads, R., Gurleyk, H. & Duxbury, J.M.: Arsenic in rice: II. Arsenic speciation in USA grain and implications for human health. *Environ. Sci. Technol.* 42 (2008), pp. 3861–3866.

Zeng, X.B., Su, S.M., Jiang, X.L., Li, L.F., Bai, L.Y. & Zhang, Y.R.: Capability of pentavalent arsenic bioaccumulation and biovolatilization of three fungal strains under laboratory conditions. *Clean-Soil Air Water* 38 (2010), pp. 238–241.

Zhang, X.H., Zhu, Y.G., Chen, B.D., Lin, A.J., Smith, S.E. & Smith, F.A.: Arbuscular mycorrhizal fungi contribute to resistance of upland rice to combined metal contamination in soil. *J. Plant Nutr.* 28 (2005), pp. 2065–2077.

Zhao, L.Y.L., Schulin, R. & Nowack, B.: Cu and Zn mobilization in soil columns percolated by different irrigation solutions. *Environ. Pollut.* 157:3 (2009), pp. 823–833.

Zhou, J.L.: Zn biosorption by *Rhizopus arrhizus* and other fungi. *Appl. Microbiol. Biotechnol.* 51 (1999), pp. 686–693.

Zhou, T., Radaev, S., Rosen, B.P. & Gatti, D.L.: Structure of the ArsA ATPase: the catalytic subunit of a heavy metal resistance pump. *Embo. J.* 19 (2000), pp. 4838–4845.

Zobrist, J., Dowdle, P.R., Davis, J.A. & Oremland, R.S.: Mobilization of arsenite by dissimilatory reduction of adsorbed arsenate. *Environ. Sci. Technol.* 34 (2000), pp. 4747–4753.

CHAPTER 7

In-situ immobilization of arsenic in the subsurface on an anthropogenic contaminated site

Timo Krüger, Hartmut M. Holländer, Jens Stummeyer, Bodo Harazim, Peter-W. Boochs & Max Billib

7.1 ARSENIC IN CHEMICAL WARFARE AGENTS

Due to its natural toxicity arsenic (As) has been used in many chemical warfare agents (CWA) during World War (WW) I and II. These CWA are on a cellular level more toxic than pure arsenic (Henriksson *et al.*, 1996). CWA were used during WW I in combat and were merely produced during WW II. These compounds are persistent to these days in soil and groundwater (Haas *et al.*, 1998). Decomposition products which are mainly analyzed in present groundwater are phenylarsonic acid, phenylarsine oxide and diphenylarsinic acid (Daus *et al.*, 2008). These decomposition products mostly do not exhibit the high toxicity of the original compounds (Schneider *et al.*, 2005). As containing substances were predominantly used in so-called "Blue Cross" CWA. The name is derived from the fact that substances with low volatility and high irritation potential were marked with a blue cross or ring (Martinetz, 1993). "Blue Cross" CWA containing As are e.g., adamsite, arsenic trichloride and Clark I.

In 1996, it was suspected that over 3200 military sites have been contaminated in the Federal Republic of Germany (Thieme, 1996). However, CWA were produced at 27 sites and had possibly been used at 132 filler points for explosives. Additionally, CWA have possibly been handled at decommissioning points inclusive blasting and burning areas (overall 480) as well as testing sites (45), ammunition factories and storages (Thieme, 1996). The necessity to remediate or at least secure the contaminated sites results from the toxicity of these compounds.

We developed and tested an *in-situ* technique for the immobilization of organic and inorganic arsenic compounds based on the subterranean deferrification and demanganesation under addition of dissolved iron which has been studied at laboratory scale before (Holländer *et al.*, 2008; Krüger *et al.*, 2008).

7.2 SITE DESCRIPTION

The groundwater at the investigated military site in Northern Germany is contaminated with As containing CWA. Other contaminants are NAPL, explosives and decomposition products of sulfurous CWA. More detailed information on the CWA is missing based on the poor documentation. The total As-concentrations (t-As) in the groundwater at the site as measured in 2005 ranges from $0.03\,\text{mg}\,\text{L}^{-1}$ to $9.05\,\text{mg}\,\text{L}^{-1}$ with arithmetic mean concentration of $2.38\,\text{mg}\,\text{L}^{-1}$ and a median concentration of $0.98\,\text{mg}\,\text{L}^{-1}$ (Holländer *et al.*, 2008). The center of the plume started near the surface (StO-01, Fig. 7.1) at one potential source of the As-contamination and lowers its depth with increasing length along the groundwater flow path. We observed the central part of the plume down to 40 m below surface along a flow length of about 130 m in this study. However, it is expected that the plume does further outreach since the observed t-As-concentration at StO-04 is still $2.40\,\text{mg}\,\text{L}^{-1}$. t-As is predominantly organically bound As (org-As) and thus mainly occurs in the form of phenylized As compounds. However, we were not able to distinguish the species of

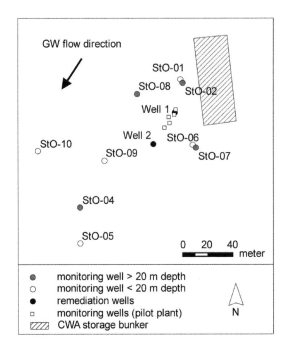

Figure 7.1.   Top view of study area.

the org-As-compounds due to their complexity directly. The inorganic As-fraction (i-As: As(III) and As(V)) ranges between 6% and 87% with a mean value of 37% and a median value of 32%. At the investigated site, during both world wars CWA were investigated, produced and tested. Additionally, CWA were disposed of using incineration and burying between and after the wars. Finally, the CWA reached soils and groundwater by this improper handling. The site as well as the sampling and measurement technique have been described in detail by Holländer *et al.* (2008) and Krüger *et al.* (2008). The remediation of the site is ongoing with a pump-and-treat facility having a capacity of $190\,m^3\,h^{-1}$. In the facility org-As is being removed by activated carbon. In the beginning of the remediation process i-As was removed by granulated ferric hydroxides (Driehaus, 2002; Driehaus *et al.*, 1998). Later, the granulated ferric hydroxide in the filters has been replaced by activated carbon as well.

We installed a pilot plant for the *in-situ* immobilization of arsenical CWA on the site. Since several hot-spots can be found, the pilot plant was installed downstream of one hot-spot. The source of the hot-spot was a former CWA storage bunker (Fig. 7.1). After WW II, the arsenic containing CWA from the bunkers cisterns were burned next to the bunker in open zinc tanks by adding kerosene. Huge amounts of CWA and their degradation products infiltrated into the soil and further into the groundwater due to tank corrosion and accumulation of arsenic in combustion residues.

A schematic cross-section of the site and the study area is shown in Figure 7.2. The sandy aquifer with a hydraulic conductivity of about $4.5 \times 10^{-4}\,m\,s^{-1}$ determined by pumping tests and a thickness of 40 m is roughly divided into two layers by marl lenses which act as local aquitards. The groundwater flow direction is from NE to SW.

The depth to the groundwater table is 5 to 7 m. The vertical markers indicate monitoring wells as well as two remediation wells (Well 1 and Well 2) which were built for the pilot plant. The t-As concentrations as measured in 2005 are given next to the well screens (dashed lines). Thus, the As plume is apparently sinking below a marl lens (aquitard) in the southern direction (Fig. 7.2).

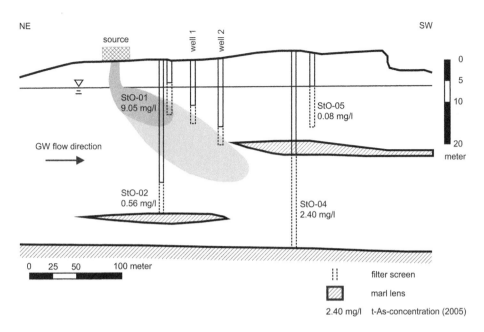

Figure 7.2.   Schematic cross-section through study area, concentrations show t-As as measured in 2005.

Table 7.1.   Geochemical parameter characterizing the site before starting the pilot trial (October 2008).

	Depth [m]	Temperature [°C]	pH [−]	EC [$\mu S\,cm^{-1}$]	Eh [mV]
StO-01	10	11.1	6.8	431	26
StO-02	21	10.0	5.0	207	65
StO-04	21	9.8	4.3	216	245
StO-06	10	10.4	6.1	227	88
StO-07	21	10.0	5.3	210	168
StO-08	21	10.7	5.8	301	19
StO-09	15	10.3	4.3	252	102
Well 1	10		7.9	196	142
Well 2	14		8.0	211	132

The site is characterized mainly by acid groundwater (pH 4.3–6.8) as measured in a field campaign from October 20–22, 2008. However, the site characterization at the two locations of the remediation well shows slightly alkaline groundwater (Table 7.1). The electrical conductivity (EC) ranges from 196 to 431 $\mu S\,cm^{-1}$. The redox conditions (Eh) is between 19 and 245 mV which mean generally either iron & manganese reducing conditions or nitrogen reducing conditions. This was verified by a campaign in 2006 where Fe(II)-concentration $<50\,\mu g\,L^{-1}$ were found while Fe(III) was not detectable (Holländer *et al.*, 2008). Furthermore, the campaign in 2008 gave for sulfate (SO_4) $42\,mg\,L^{-1}$, for phosphate (PO_4) about $0\,mg\,L^{-1}$, and for silica (SiO_2) $3\,mg\,L^{-1}$.

Both remediation wells are bidirectional wells. For observation purposes, five observation wells for sampling and monitoring of water parameters and levels have been installed (Fig. 7.1).

The remediation wells have been installed downstream of the CWA storage bunker having a depth of 15 m (well 1) and 19 m (well 2) with a screen length of 4 m each. The distance between the wells is 30 m. The pumps have been installed 2 m above the filter screens. In order to reduce the risk of iron incrustations at the pumps, additional infiltration pipes for Fe-enriched water were installed ending one meter below the pumps. Thus, the wells can both pump and infiltrate water. t-As-concentration of 1.3 mg L^{-1} and 2.5 mg L^{-1} were measured at well 1 and in well 2, respectively after installing the wells.

7.3   REMEDIATION METHOD

7.3.1   *Precipitation and sorption by metals*

Since i-As adsorbs to iron oxides (Chapter 3) it is possible to reduce the concentration in the aquatic phase by adding iron salts. The removal of i-As from water has been thoroughly investigated in batch experiments (Bissen and Frimmel, 2003; Dixit and Hering, 2003; Hering et al., 1996; Pierce and Moore, 1982; Rott and Meyer, 2000; Scott et al., 1995; Violante et al., 2006; 2007; Wilkie and Hering, 1996). It could be demonstrated that iron (Fe) was suitable to remove i-As from a solution. The needed amount of applied Fe was always greater than the amount of As in the solution (8:1–40:1). The adsorption of i-As is strongly influenced by availability of competitors for sorption spots such as sulfate, phosphate, silica, ammonia and bicarbonate. Thus the efficiency of the remediation method is dependent on the hydrogeochemical environment. However, the concentrations of potential competitors were rather low (SO$_4$ 42 mg L^{-1}, PO$_4$ ~0 mg L^{-1}, and SiO$_2$ 3 mg L^{-1}). Holländer et al. (2008) and Krüger et al. (2008) confirmed the literature results with the ambient groundwater from the military site. They removed more than 90% of i-As.

Daus et al. (2008) found that the CWA decomposition product phenylarsonic acid was removed from a solution by precipitating Fe. Holländer et al. (2008) and Krüger et al. (2008) showed that organic arsenicals from CWA were precipitated by adding different Fe compounds. They added 50 mg zero-, bi- or trivalent Fe to 250 mL of water containing 9 mg L^{-1} t-As (66% org-As) and precipitated the iron by aeration. Under those conditions, they could demonstrate that the t-As-concentration can be reduced in all experiments. However, bivalent ferrous chloride (FeCl$_2$) and ferrous sulfate (FeSO$_4$) showed the best efficiency of about 80% in removal of t-As. Further batch experiments showed that adding FeCl$_2$ to solutions at different As compositions and concentrations resulted in removal efficiencies over 90% for i-As and 50–75% for org-As. Additional soil column experiments showed that 1 g t-As could be removed using of 5 to 20 g Fe (Krüger et al., 2008).

7.3.2   *Remediation technique*

Subterranean deferrification and demanganesation (Rott et al., 1996) has proven to decrease As-concentrations in treated water. While oxidizing the Fe compounds, *in-situ* precipitated Fe hydroxide coats the grain surface in the subsurface. As adsorbs on these Fe compounds and thus is removed from the liquid phase (Rott and Meyer, 2000; Rott et al., 1996). Various experiments showed the possibility to remove As cost-efficiently by subterranean deferrification and demanganesation (Kauffmann, 2008; Rott and Meyer, 2000; Rott et al., 1996; van Halem et al., 2010). Since the removal of As always requires a higher concentration of Fe(II) than As, this method cannot be applied at sites where the Fe concentration is too low. Thus the subterranean deferrification and demanganesation has been enhanced by an additional Fe dosage to overcome these limitations. Ferric incrustations are an emerging problem for this technique if the bivalent Fe reacts especially within the well with dissolved oxygen.

For building an effective subterranean reaction zone, the resulting pilot plant for the military site comprises two bidirectional wells and a container with measurement and control technology.

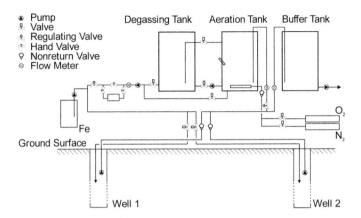

Figure 7.3.   Technical design of the pilot plant.

Three tanks have been installed in this container (Fig. 7.3): the first tank (aeration tank, 500 L) allows increasing or decreasing of the concentration of dissolved oxygen. Excess gas bubbles are removed from the infiltrated water in the second tank (degassing tank, 500 L). The third tank (buffer tank, 500 L) receives the part of the pumped water which is afterwards pumped to the remediation facility. $FeCl_2$ solution can be added into the degassed water which is finally infiltrated into the infiltration well. The infiltration pipe is equipped with a bypass for measuring in situ water quality parameters such as pH, redox potential, dissolved oxygen (DO) and electric conductivity (EC).

The two wells have a very short screen length of 4 m. Short filter screens are very important since they reduce the dwell period of the Fe enriched water and therefore reduce the risk of Fe incrustations. Additionally, the wells are equipped with two pipes so that no infiltration water uses the pipes of the pumped water. The infiltration pipe reaches down to the filter screen. Thus, infiltrated Fe enriched water has only a short, direct contact to the potentially oxygen enriched water within the well and is delivered to the aquifer more directly. As a result of the short screen length, low maximum pumping and infiltration rate of $6\,m^3\,h^{-1}$ had to be applied.

The schedule of the pilot plant is characterized by the repetition of similar cycles, which generally differ in the pumping direction. One cycle is subdivided into four steps (Fig. 7.4).

Step (1) – infiltrating Fe-enriched water: Water is pumped into the aeration tank from well 1 and stripped with nitrogen gas (N_2) to decrease the DO-concentration in the water. The water is infiltrated into well 2 after amending the water with $FeCl_2$-solution once the DO-concentration drops below $0.5\,mg\,L^{-1}$. The applied $FeCl_2$-solution has a Fe content of 118 g per liter $FeCl_2$-solution.

Step (2) – infiltrating water with low DO-concentration: Water is pumped from well 1 through the aeration and the degassing tank and stripped with nitrogen in the aeration tank. During this step, the groundwater with dissolved Fe is forced to flow from the well into the aquifer.

Step (3) – infiltrating water with a high DO-concentration: This step is similar to step 2 with the difference that ambient air instead of nitrogen is dosed in the aeration tank. In this step groundwater with high dissolved oxygen concentration is transported into the aquifer. DO reacts with the dissolved Fe and therefore the Fe precipitates.

Step (4) – infiltrating untreated water: Water is pumped from well 1 through the aeration tank as well as the degassing tank and directly infiltrated into well 2 without further treatment. This step creates a buffer around well 2 in which water with a lower concentration of DO is stored than infiltrated in step (3). Applying this method the risk of clogging within the well is reduced because the pumping direction is switched on the next day and water is pumped from well 2 and then again stripped with N_2 [step (1)].

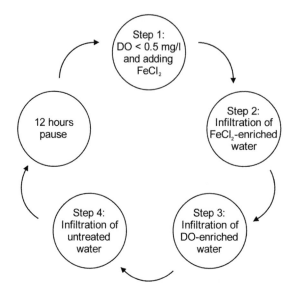

Figure 7.4.   Work flow for the pilot plant operation.

The above described steps comprise one cycle (Fig. 7.4). The cycles are completely computerized as well as the internal control and measurements within the pilot plant. For the second cycle, all steps are repeated with the difference that water is pumped from well 2 and infiltrated into well 1. Thus a reactive domain around both wells is established due to the forced precipitation of Fe. A part of the pumped water is always discharged to the external remediation facility during the pilot plant operation. This part reached up to 20% (1 m³ h⁻¹) of the withdrawn discharge during the pilot experiment. Thus, a higher amount of water is withdrawn than infiltrated. Therefore, water from outside of the subterranean reactive domain will enter the domain, and the As of this water is removed by adsorption. The pump-, infiltration- and discharge rates for the steps throughout the experiment can be seen in Table 7.2. As the main result, the As-concentration of the water which is pumped to the external remediation plant is lower than in the ambient groundwater.

A field study was carried out over a period of 705 days for the infiltration of Fe. In the beginning 50 mL FeCl₂ solution containing 5.9 g Fe has been dosed per cycle (11.8 mg Fe L⁻¹ infiltrated water). The dosage was increased stepwise up to 1000 mL (118 g Fe) per cycle (236 mg Fe L⁻¹). The amount was increased when there was no or little Fe measured in the water pumped from the wells. The low dosage in the beginning was chosen to achieve the first aim of the pilot study: Immobilization of the i-As as shown in the batch experiments targeting on a rate of 10 g Fe to 1 g i-As. The second reason for choosing the low dosage was to avoid clogging of the pores near the wells or even at the well screens. An indicator is dissolved Fe in the pumped water. If Fe were measured in the pumped water, not all of the infiltrated Fe was precipitated in the aquifer. The second indicator was the immobilization of org-As. Therefore, the Fe dosage was increased throughout the study.

The amount of dosed Fe is shown in Table 7.2. Additionally the pump and infiltration rates are given. A typical schedule for one day is depicted in (Fig. 7.4).

The steps have always been carried out in the sequences (1), (2), (3) and (4). Every entry in Table 7.2 marks a change in the schedule. If a step is not listed in the table, the parameters for the step remained unchanged. In the beginning, all steps had a pump rate of 3.0 m³ h⁻¹ and discharge water of 0.5 m³ h⁻¹. The infiltration rate was 1.0 m³ h⁻¹ in step (1) and 2.5 m³ h⁻¹ in all other steps. The durations of the steps were 0.5 h, 5 h and 2 h for steps (2), (3) and (4). Step (1) had

Table 7.2. Parameter for the steps of the pilot study; horizontal dashed lines mark a change in the schedule, steps not shown after a change in schedule remain unchanged.

Time [d]	Step	Pumping rate [m³ h⁻¹]	Infiltration rate [m³ h⁻¹]	Discharge rate [m³ h⁻¹]	FeCl₂-dosage [mL/cycle]	Fe-dosage [g/cycle]	Fe-dosage [mg L⁻¹]	Duration [h]
0–22	1	3.0	1.0	0.5	50	5.9	11.8	–
	2	3.0	2.5	0.5	–	–	–	0.5
	3	3.0	2.5	0.5	–	–	–	5.0
	4	3.0	2.5	0.5	–	–	–	2.0
23–40	1	3.0	1.0	0.5	100	11.8	23.6	–
41–103	1	3.0	1.0	0.5	150	17.7	35.4	–
104–134	1	3.0	1.0	0.5	200	23.6	47.2	–
135–194	1	3.0	1.0	0.5	300	35.4	70.8	–
195–222	1	3.0	1.0	0.5	375	44.3	88.5	–
223–333	1	3.0	1.0	0.5	500	59.0	118.0	–
334–355	1	3.5	1.0	1.0	500	59.0	118.0	–
	2	3.5	2.5	1.0	–	–	–	0.5
	3	3.5	2.5	1.0	–	–	–	5.0
	4	3.5	2.5	1.0	–	–	–	2.0
356–377	1	2.8	1.0	0.3	500	59.0	118.0	–
	2	2.8	2.5	0.3	–	–	–	0.5
	4	2.8	2.5	0.3	–	–	–	2.0
378–466	1	2.8	1.0	0.3	750	88.5	177.0	–
467–487	1	2.8	1.0	0.3	1000	118.0	236.0	–
488–501	1	–	–	–	–	–	–	–
	2	–	–	–	–	–	–	–
	3	3.0	2.5	0.5	–	–	–	7.0
	4	–	–	–	–	–	–	–
502–533	3	–	–	–	–	–	–	–
	4	3.0	2.5	0.5	–	–	–	7.0
534–637	1	2.8	1.0	0.3	1000	118.0	236.0	–
	2	2.8	2.5	0.3	–	–	–	0.5
	3	2.8	2.5	0.3	–	–	–	5.0
	4	2.8	2.5	0.3	–	–	–	2.0
638–705	1	4.1	1.0	0.3	1000	118.0	236.0	–
	2	4.1	3.9	0.3	–	–	–	0.5
	3	4.1	3.9	0.3	–	–	–	5.0
	4	4.1	3.9	0.3	–	–	–	2.0

no fixed duration since reaching the concentration threshold for oxygen lasted a variable amount of time.

The dosage of FeCl₂-solution was increased in six stages up to 500 mL/cycle (118 mg Fe L⁻¹). The discharge water was increased to 1 m³ h⁻¹ after 334 days and reduced for the steps (1), (2) and (4) after 365 days again. The next changes were the increase of the dosage up to 1000 mL/cycle (238 mg Fe L⁻¹). Subsequently, a remobilizing experiment was conducted between day 488 and day 533. Hence, the FeCl₂ dosage was switched off, and water with an increased DO-concentration was infiltrated into the wells for two weeks after which the oxygen enrichment was switched off for another four weeks.

In the last phase, it was attempted to decrease the t-As concentration to a technical minimum. For this, the discharge rate was decreased to 0.3 m³ h⁻¹ (operational minimum of the pilot plant) and the Fe dosage was increased to 1000 mL/cycle (236 mg Fe L⁻¹, operational maximum of the pilot plant). After 103 days (day 638 of the experiment) the pumping rate was increased from 3.0

to $4.1 \, m^3 \, h^{-1}$ (operational maximum of the pilot plant with high DO-saturation) with unchanged discharge. Therefore, the amount of process water was increased and thus the reactive domain in the aquifer. Van Halem *et al.* (2010) stated that the efficiency of the As-immobilization is improved by this procedure.

7.4   FIELD EXPERIMENT RESULTS

7.4.1   *Arsenic concentration*

The analysis results from the infiltration experiment are depicted in the following diagrams. The sampling and analysis methods have been described in Stummeyer *et al.* (1996) and Holländer *et al.* (2008). Circles and triangles mark the mean value of all 16 measurements during one cycle. Triangles indicate results from well 1, circles from well 2. The error bars mark the minimum and maximum value within a cycle. The black bars at the bottom show times when the plant was not working properly or shut down. The dosage of the $FeCl_2$ solution is shown as a bar diagram in the background. Changes in pump rate or dosage are marked by vertical dashed lines.

Figure 7.5 shows the t-As concentration throughout the entire experiment. Generally, the t-As concentration decreases over the runtime although no significant reduction has been observed at a dosage of 50 mL/cycle. The mean concentration in well 1 is lower than in well 2 in the beginning of the study. This is most likely due to the position of the well screens. They were built in different depths and thus most likely different environments. The increase of the dosage to 100 mL/cycle resulted in a significant decrease in the t-As concentration. The concentration decreased to $1 \, mg \, L^{-1}$ t-As at the $FeCl_2$-dosage of 150 mL/cycle. After a phase of limited plant functionality, the t-As concentration is temporarily increased but decreased again when the plant was switched on again. No significant trend in the t-As concentration was noticed at 200 mL/cycle.

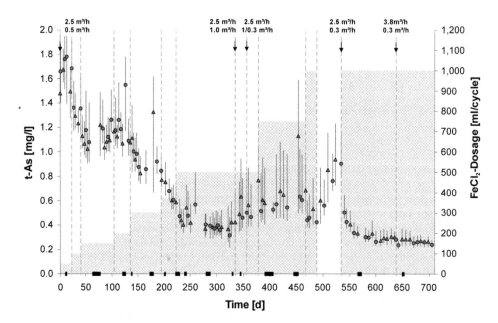

Figure 7.5.   Mean values of t-As-concentration per cycle in well 1 (triangles) and well 2 (circles); error bars mark minimum and maximum of values in a cycle, gray bars denote the $FeCl_2$-dosage.

This might be due to problems occurred with the Fe dosage during this phase so that potentially a lower amount of Fe was added. Increasing the dosage to 300, 375 and 500 mL decreased the t-As concentration below $0.4 \, \text{mg} \, \text{L}^{-1}$.

Noticeably, the mean concentration in well 2 started out higher than the one in well 1 and remained higher for about 200 days of the experiment. After 200 days almost no difference could be noticed between the As concentrations in the two wells. Finally, the concentration in well 2 was lower than in well 1 after day 330. This was estimated earlier since well 2 received the downstream water from well 1 which has already decreased As-concentration.

Changing the amount of discharged water to the pump-and-treat facility from 0.5 to $1.0 \, \text{m}^3 \, \text{h}^{-1}$ on day 334 leads to an increase in t-As concentration up to $0.8 \, \text{mg} \, \text{L}^{-1}$. An increase in concentration was expected since the amount of injected Fe was constant while the flow of ambient (contaminated) groundwater into the subterranean reactor was larger due to the higher discharge which increased by $0.5 \, \text{m}^3 \, \text{h}^{-1}$. The deviation of the minimum and maximum values from the mean value of a cycle was increased as well (Fig. 7.5). An increase in dosage to 750 mL and 1000 mL decreased the mean concentration to $0.4 \, \text{mg} \, \text{L}^{-1}$ again.

The immobilization phase was followed by a 45 days remobilization phase. After switching the Fe dosage off, the t-As concentration increased by $0.5 \, \text{mg} \, \text{L}^{-1}$ which is an increase of more than 100%. Nevertheless, the concentration were below $1 \, \text{mg} \, \text{L}^{-1}$ and thus much lower than the initial concentration of about $1.6–1.7 \, \text{mg} \, \text{L}^{-1}$. Hydraulic observations showed an effective flow velocity of $1 \, \text{m} \, \text{d}^{-1}$. Therefore, the groundwater flows about 45 m within the 45 days remobilization period. The distance between the two wells is 30 m so that the subterranean reactor was still able to immobilize As because the resulting concentrations were still much lower than the initial concentration which is still flowing into the system from upstream. Hence the precipitated and adsorbed As had not been remobilized.

The cumulative graph of the immobilized t-As is presented in Figure 7.6. The data were calculated by using the mean As-concentration of each dosage period. The immobilized mass of t-As can be calculated by taking the amount of discharged water to the pump-and-treat facility

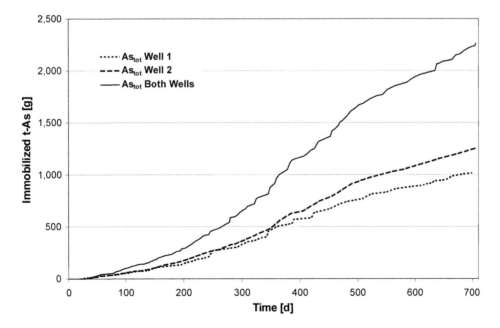

Figure 7.6.   Cumulative graph of immobilized t-As.

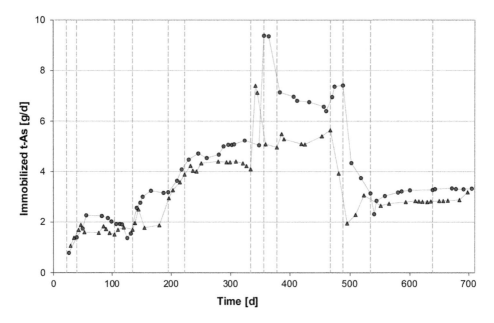

Figure 7.7.   Immobilized t-As per day; triangles denote concentrations in well 1, circles in well 2.

and its t-As concentration into account. The initial concentrations from both wells were used for all time steps since no measurement upstream of the wells and outside the reactive domain were available. We assumed the initial concentration as a constant for the inflow concentration although there might have been potential changes of the inflow concentration. However, since the As-concentrations were observed for a couple of years by the management of the pump-and-treat-facility and have not reduced significantly, the estimated error in this regard is assumed to be rather small compared to the total amount of As in the system.

Well 2 shows a higher immobilized mass than well 1 due to its higher initial concentration. The cumulative graph shows a flat slope in the beginning since low $FeCl_2$ dosage results only in low amounts of immobilized As. The slope gets steeper after about 250 days (500 mL/cycle). At the end of the pilot study about 1.0 kg of t-As was immobilized near well 1 and 1.3 kg near well 2 respectively. Thus adding up to a total of 2.3 kg of immobilized As.

The immobilization rate of t-As is shown in Figure 7.7. The immobilization rate increased with rising Fe dosage. The rate increased significantly when the discharge rate was increased from 0.5 to 1.0 m^3 h^{-1}. This was due to doubling the amount of water in the calculation but still having similar concentrations. The processes in and around the subsurface reactive area reach stable conditions after around 30–50 days of unchanged operating parameters and thus are rather slow. Since the discharged water was decreased again from 1.0 to 0.3 m^3 h^{-1} at day 356 again, the immobilization rate 'decreased' as well so that the operating period was too short to reach stable conditions. At a discharge of 5.75 m^3 d^{-1} and a $FeCl_2$-dosage of 750 mg L^{-1} an immobilization rate of 6 g d^{-1} was achieved.

The ratio of dosed Fe to immobilized t-As is shown in Figure 7.8. The lower the ratio the lesser Fe is needed to immobilize As and the more efficiently Fe is used. The ratio stayed constant between 10:1 and 15:1 up to a dosage of 750 mL $FeCl_2$/cycle. At the beginning of the experiment the ratio was below 10:1 and increased at a dosage of 200 mL/cycle up to almost 20:1 in well 1. The low ratio can be explained by As-compounds (mainly inorganic) which adsorb very easily at the iron hydroxide (see Section 7.4.2). As more of this As has already adsorbed the larger was the ratio with the on-going duration of the pilot study. The mentioned problems concerning the

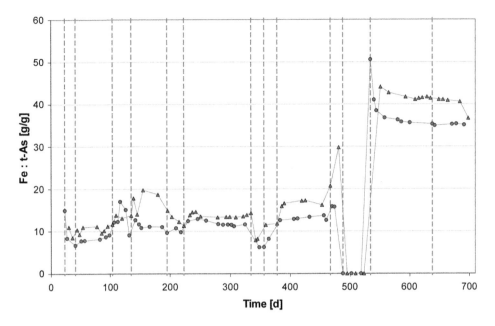

Figure 7.8.   Ratio of dosed Fe to immobilized t-As; triangles denote well 1, circles well 2.

$FeCl_2$-dosage system are one explanation of the high ratio of 20:1. After increasing the discharge from 0.5 to 1.0 m^3 h^{-1} the ratio decreases significantly to about 8:1. This is an analogue to the calculated rate of the immobilized t-As which was due to the non-stable conditions (Fig. 7.7). The attempt to reach the minimum t-As concentration with high Fe dosage and low discharge showed, as expected, a high ratio (35:1 to 45:1). This shows that dosing low amounts of $FeCl_2$ is efficient in regard of the Fe:As ratio but dosing high amounts of $FeCl_2$ is efficient in regard of the amount of removed As.

7.4.2   *Change in arsenic species distribution*

During the field experiment a speciation of t-As into i-As and org-As has been done following a method described in detail in Krüger *et al.* (2008) and Holländer *et al.* (2008). At the beginning of the experiment the fraction of i-As amounted to 11–12% in well 1 and 13–16% in well 2. With increasing dosage the percentage of i-As decreases. The fraction of i-As in t-As decreased below 10% during the first three dosing periods. During a break in the plant operation the fraction increased again. The concentration of i-As decreased temporarily below the detection limit at a dosage of 500 mL/cycle and 0.5 m^3 h^{-1} discharge. Considering the entire experiment an increase can be noticed only during the remobilization experiment where the i-As fraction increased from 1.5% to 5%. During the last two dosing periods the concentration decreased below the detection limit again resulting in an i-As fraction near 0%. The zero value was not reached as the detection limit was used as the lowest limit in the calculations.

The decrease in the fraction of i-As in t-As implies that i-As is preferably adsorbed to the iron hydroxide. While the t-As-concentration decreases from 100 to 18% of the initial concentration, the i-As-concentration decreases from 100% to less than 1%. This was already observed during the laboratory experiments (Holländer *et al.*, 2008). The bulk of the immobilized t-As is org-As due to its larger amount in the ambient groundwater but i-As is preferentially adsorbed to the Fe.

When comparing the actual concentration of i-As and org-As to their respective initial concentrations as done in Figure 7.9 it can be seen that the relative concentration of i-As quickly

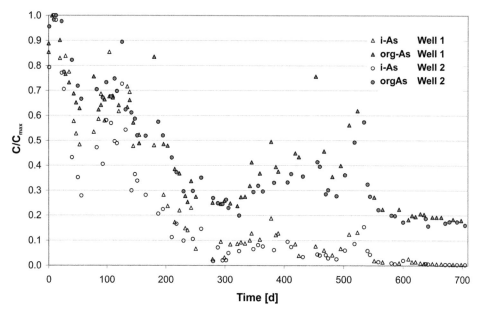

Figure 7.9.   Comparing C/C_{max} of org-As (grey symbols) and i-As (white symbols) in well 1 (triangles) and well 2 (circles).

falls below the relative concentration of org-As. At the end of the field experiment the relative concentration of i-As is 1% and 18% for org-As.

7.4.3   *Iron concentration*

The Fe concentration in the pumped water is critical for the life expectancy of the remediation wells. Due to potential clogging problems it is necessary that the Fe concentration is as low as possible. The pumped water has a mean concentration $<100\,\mu g\,L^{-1}$ (Fig. 7.10). Higher concentrations have been observed during the phase when the plant did not operate correctly. The concentration reduced directly after the correct schedule of the plant. This can be explained by the fact that during this period Fe was infiltrated but only a small amount of water with enriched oxygen was infiltrated. In the absence of oxygen Fe is inhibited from precipitating.

Generally a high fraction of Fe was precipitated: Assuming a uniform distribution of Fe in the pumped water, a $3\,m^3\,h^{-1}$ pumping rate and a 50 mL/cycle dosage the mean concentration in the water would be $0.26\,mg\,L^{-1}$. At $2.8\,m^3\,h^{-1}$ pumping rate and 1.000 mL/cycle dosage the mean concentration would be $5.7\,mg\,L^{-1}$. Since the mean Fe concentration that was measured in the groundwater is significantly below these values it can be assumed that the dosed Fe is reliably precipitated in the aquifer.

Figure 7.10 shows that Fe was not detectable during the remobilizing experiment between day 488 and day 533. This behavior was estimated in the first two weeks (day 488 to 501) since DO-enriched water was infiltrated. However, the Fe-concentration stayed below detectable limits for the rest of the remobilization experiment after shut-down of this dosing (day 502 to 533). This indicates that the iron hydroxides complexes at which the As is adsorbed are stable and agrees with literature. E.g., Henning (2004) showed that the initially voluminous iron hydroxides complexes are dewatered to more compact forms like hematite while they become older. The compact forms are much more stable compared to the voluminous iron hydroxides complexes.

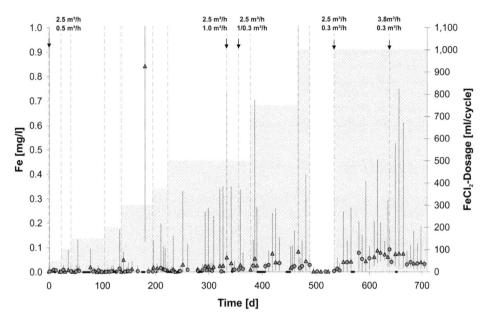

Figure 7.10.   Mean values of Fe concentration per cycle in well 1 (triangles) and well 2 (circles); error bars denote the minimum and maximum of values in a cycle.

7.5 CONCLUSIONS

We presented in this chapter an *in-situ* method for the immobilization of arsenic from chemical warfare agents. The method is based on subterranean deferrification and demanganesation with an artificial infiltration of bivalent Fe. We showed that inorganic as well as chemical warfare agent based (mainly organic) arsenic compounds could be immobilized *in-situ*. During the pilot study which lasted almost two years, the initial arsenic concentration of around $1.65\,\mathrm{mg\,L^{-1}}$ in the groundwater has been reduced to less than $0.3\,\mathrm{mg\,L^{-1}}$ and in total more than 2.3 kg As have been immobilized.

The main advantage of the method is that no contaminated sludge or adsorption material has to be disposed off. One of the greatest advantages of the method is that all operating cycles use the same pore space. Therefore, the contaminated groundwater has to flow through the reactive domain before it reaches the well and the infiltrated Fe has to have contact with the infiltrated oxygen. Although this describes also the biggest challenges when applying the method: the iron hydroxide has to precipitate in the pores, and there is a risk that iron hydroxide precipitates even at the screen or in the well. This would induce clogging and has to be prevented. We did so by stripping the water with nitrogen. Another possibility is to decrease the pH-value of the water before infiltration. As a result, we did not observe any hydraulic response in means of lower infiltration capacities during our study although a camera survey showed that a soft film covered the well screens after one year of operation. Finally, in the efficiency test of the wells we determined that the maximum pump rate was not decreased significantly in the wells.

The hydrogeochemical environment of the application site has to be evaluated before applying this method. High concentrations of competitors for sorption spots may decrease the efficiency of the method and highly reducing environment may lead to the iron being remobilized and thus the As, as well. However, we have not found any evidence of remobilizing iron during the remobilizing experiment at the field site.

Further to our case, the method can be applied for other case studies where the natural Fe concentration is too low to adsorb all arsenic species. For example, the method can be applied for sustainable drinking water production or the remediation of a contaminated site. Drinking water production of course should not be done on a site contaminated with chemical warfare agents but rather in regions with a naturally increased As concentration. In addition, it has been reported by various groups that heavy metals can be adsorbed to Fe compounds (Ball and Nordstrom, 1991; Buekers *et al.*, 2008; Dzombak and Morel, 1990; Gerth and Brümmer, 1983). This shows that the application of this method is not only restricted to the removal of As from water.

ACKNOWLEDGEMENTS

These investigations were supported by the Federal Ministry of Defense (BMVg) and the Construction Department of Lower Saxony – Federal Competence Centre for Soil and Groundwater Protection. The authors also thank the two anonymous reviewers for valuable comments that improved the quality of this article.

REFERENCES

Ball, J.W. & Nordstrom, D.K.: User's manual for WATEQ4F, with revised thermodynamic data base and test cases for calculating speciation of major, trace, and redox elements in natural waters. US Geological Survey Open File report 91-183. US Geological Survey, Menlo Park, CA, 1991.

Bissen, M. & Frimmel, F.H.: Arsenic — a review. Part II: oxidation of arsenic and its removal in water treatment. *Acta Hydrochim. Hydrobiol.* 31:2 (2003), pp. 97–107.

Buekers, J., Amery, F., Maes, A. & Smolders, E.: Long-term reactions of Ni, Zn and Cd with iron oxyhydroxides depend on crystallinity and structure and on metal concentrations. *Eur. J. Soil Sci.* 59:4 (2008), pp. 706–715.

Daus, B., Mattusch, J., Wennrich, R. & Weiss, H.: Analytical investigations of phenyl arsenicals in groundwater. *Talanta* 75:2 (2008), pp. 376–379.

Dixit, S. & Hering, J.G.: Comparison of arsenic(V) and arsenic(III) sorption onto iron oxide minerals: implications for arsenic mobility. *Environ. Sci. Technol.* 37:18 (2003), pp. 4182–4189.

Driehaus, W.: Arsenic removal – experience with the GEH® process in Germany. *Water Sci. Technol. Water Supply* 2:2 (2002), pp. 275–280.

Driehaus, W., Jekel, M. & Hildebrandt, U.: Granular ferric hydroxide—a new adsorbent for the removal of arsenic from natural water. *J. Water Supply Res. Technol.-Aqua* 47 (1998), pp. 30–35.

Dzombak, D.A. & Morel, F.M.M.: *Surface complexation modeling hydrous ferric oxide*. John Wiley & Sons Inc, New York, NY, 1990.

Gerth, J. & Brümmer, G.: Adsorption und Festlegung von Nickel, Zink und Cadmium durch Goethit (a-FeOOH). *Fresen. J. Anal. Chem.* 316:6 (1983), pp. 616–620.

Haas, R., Krippendorf, A., Schmidt, T., Steinbach, K. & von Löw, E.: Chemisch-analytische Untersuchung von Arsenkampfstoffen und ihren Metaboliten. *Umweltwissenschaften und Schadstoff-Forschung* 10:5 (1998), pp. 289–293.

Henning, A.-K.: *Biologische Mechanismen bei der unterirdischen Aufbereitung von Grundwasser am Beispiel des Mangans*. Oldenbourg Industrieverlag GmbH, Stuttgarter Berichte zur Siedlungswasserwirtschaft 176, Stuttgart, Germany, 2004.

Henriksson, J., Johannisson, A., Bergqvist, P.A. & Norrgren, L.: The toxicity of organoarsenic-based warfare agents: in vitro and in vivo studies. *Arch. Environ. Con. Tox.* 30:2 (1996), pp. 213–219.

Hering, J.G., Chen, P.-Y., Wilkie, J.A., Elimelech, M. & Liang, S.: Arsenic removal by ferric chloride. *American Water Works Association, e-journal* 88:4 (1996), pp. 155–167.

Holländer, H.M., Stummeyer, J., Harazim, B., Boochs, P.-W., Billib, M. & Krüger, T.: Subsurface treatment of arsenic in groundwater – experiments at laboratory scale. In: J. Bundschuh, P. Armienta, P. Birkle, P. Bhattacharya, J. Matschullat & A.B. Mukherjee (eds): *Natural arsenic in groundwaters of Latin America*. Taylor & Francis, London, UK, 2008, pp. 537–545.

Kauffmann, H.: *Arsenelimination aus Grundwasser*. PhD Thesis, University of Stuttgart, Stuttgart, Germany, 2008.

Kopecz, P. & Thieme, J.: *Bestandsaufnahme von Rüstungsaltlastenverdachtsstandorten in der Bundesrepublik Deutschland. Vol. 3: Kampfstofflexikon.* Umweltbundesamt, Berlin, Germany, 1996.

Krüger, T., Holländer, H.M., Boochs, P.-W., Billib, M., Stummeyer, J. & Harazim, B.: In situ remediation of arsenic at a highly contaminated site in Northern Germany. In M.G. Trefry (ed): *Groundwater quality: Securing groundwater quality in urban and industrial environments.* IAHS 324, 2008, pp. 118–125.

Martinetz, D.: Arsenorganische Verbindungen in Rüstungsaltlasten. *TerraTech* 4 (1993), pp. 38–40.

Pierce, M.L. & Moore, C.B.: Adsorption of arsenite and arsenate on amorphous iron hydroxide. *Water Res.* 16:7 (1982), pp. 1247–1253.

Rott, U. & Meyer, C.: Die unterirdische Trinkwasseraufbereitung – ein Verfahren zur rückstandsfreien Entfernung von Arsen. *Wasser und Abfall* 2000:10 (2000), pp. 36–42.

Rott, U., Meyerhoff, R. & Bauer, T.: In situ-Aufbereitung von Grundwasser mit erhöhtem Eisen-, Mangan-, und Arsengehalten. *GWF* 96:7 (1996), pp. 358–363.

Schneider, K., Hassauer, M., Akkan, Z., Gfatter, S. & Oltmanns, J.: Ableitung von Prüfwerten für Kampfstoffe und Abbauprodukte für die Wirkungspfade Boden-Mensch (direkter Kontakt) und Boden-Gewässer. Report, Forschungs- und Beratungsinstitut Gefahrstoffe, FoBiG GmbH, Freiburg, i. Br., Germany, 2005.

Scott, K.N., Green, J.F., Do, H.D. & McLean, J.S.: Arsenic removal by coagulation. *American Water Works Association J. AWWA* 87:4 (1995), pp. 114–126.

Stummeyer, J., Harazim, B. and Wippermann, T.: Speciation of arsenic in water samples by high-performance liquid chromatography-hydride generation-atomic adsorption spectrometry at trace levels using a post column reaction system. Fresenius' Journal of Analytical Chemistry 3543 (1996), pp. 344–451.

Thieme, J.: *Bestandsaufnahme von Rüstungsaltlastenverdachtsstandorten in der Bundesrepublik Deutschland. Vol. 1: Bericht.* Umweltbundesamt, Berlin, Germany, 1996.

van Halem, D., Olivero, S., de Vet, W.W.J.M., Verberk, J.Q.J.C., Amy, G.L. & van Dijk, J.C.: Subsurface iron and arsenic removal for shallow tube well drinking water supply in rural Bangladesh. *Water Res.* 44:19 (2010), pp. 5761–5769.

Violante, A., Ricciardella, M., Del Gaudio, S. & Pigna, M.: Coprecipitation of arsenate with metal oxides: nature, mineralogy, and reactivity of aluminum precipitates. *Environ. Sci. Technol.* 40:16 (2006), pp. 4961–4967.

Violante, A., Gaudio, S.D., Pigna, M., Ricciardella, M. & Banerjee, D.: Coprecipitation of arsenate with metal oxides. 2. nature, mineralogy, and reactivity of iron(III) precipitates. *Environ. Sci. Technol.* 41:24 (2007), pp. 8275–8280.

Wilkie, J.A. & Hering, J.G.: Adsorption of arsenic onto hydrous ferric oxide: effects of adsorbate/adsorbent ratios and co-occurring solutes. *Colloids Surfaces* A 107 (1996), pp. 97–110.

Subject index

Arsenic in the Environment

Book Series Editor: Jochen Bundschuh & Prosun Bhattacharya

ISSN: 1876-6218

Publisher: CRC Press/Balkema, Taylor & Francis Group

1. Natural Arsenic in Groundwaters of Latin America
   Jochen Bundschuh, M.A. Armienta, Peter Birkle, Prosun Bhattacharya,
   Jörg Matschullat & A.B. Mukherjee (eds)
   2009
   ISBN: 978-0-415-40771-7 (Hbk)

2. The Global Arsenic Problem: Challenges for Safe Water Production
   Nalan Kabay, Jochen Bundschuh, Bruce Hendry, Marek Bryjak, Kazuharu Yoshizuka,
   Prosun Bhattacharya & Süer Anaç (eds)
   2010
   ISBN: 978-0-415-57521-8 (Hbk)

3. The Taiwan Crisis: A Showcase of the Global Arsenic Problem
   J.-S. Jean, J. Bundschuh, C.-J. Chen, H.-R. Guo, C.-W. Liu, T.-F. Lin & Y.-H. Chen
   2010
   ISBN: 978-0-415-58510-1 (Hbk)

4. Arsenic: Natural and Anthropogenic
   E. Deschamps & J. Matschullat (eds)
   2011
   ISBN: 978-0-415-54928-8 (Hbk)

5. The Metabolism of Arsenite
   Joanne M. Santini & Seamus A. Ward (eds)
   2012
   ISBN: 978-0-415-69719-4 (Hbk)